Music and Acoustics

Music and Acoustics

From Instrument to Computer

Philippe Guillaume

Series Editor
Société Française d'Acoustique

iSTE

First published in France by Hermes Science/Lavoisier entitled "Musique et acoustique : de l'instrument a l' ordinateur"
First published in Great Britain and the United States by ISTE Ltd

ISTE Ltd
6 Fitzroy Square
London W1T 5DX
UK
www.iste.co.uk

ISTE USA
4308 Patrice Road
Newport Beach, CA 92663
USA

First South Asian Edition 2007

Library of Congress Cataloging-in-Publication Data

Guillaume, Philippe, 1955-
[Musique et acoustique. English]
Music and acoustics : from instrument to computer/Philippe Guillaume.
p. cm.
Includes bibliographical references (p.) and index.
ISBN 13: 978-1-905209-26-6
ISBN 10: 1-905209-26-6
1. Music--Acoustics and physics. 2. Computer sound processing. I. Title.
ML3805,G8513 2006
781.2'3—dc22

2006028748

British Library Cataloguing-in-Publication Data
A CIP record for this book is available from the British Library
ISBN 10: 1-905209-26-6
ISBN 13: 978-1-905209-26-2

Printed in Brijbasi Art Press Ltd., I-72, Sector-9, Noida, U.P. India.

To Nicole, Cécile and Olivier

Contents

Foreword

How does a tuner achieve such a precise tuning of a piano or an organ using nothing but his ears? Why does the clarinet, though equal in length to the C flute, play one octave lower? What difference is there between the Pythagorean scale and the tempered scale? How can a series of notes seem to rise indefinitely even though it always repeats the same notes? What are the possibilities offered by digital sound? What are its limitations? How can a compression technology such as MP3 achieve a tenfold reduction of a sound file's size without significantly altering it? What is the very simple principle underlying audio synthesis in Yamaha's famous keyboard, the DX7? These are a few examples of the questions we will try to answer.

The goal of this book is to use these questions to give the reader an overview of the nature of musical sound, from its production by traditional musical instruments to sounds obtained by audio synthesis, without trying to be exhaustive however: this book is not meant as a catalogue, but instead, I hope, as a first step that will enable the reader to move on to more specific areas in this field. Musical sound is addressed from a scientific standpoint, and the succession of causes that lead to a specific type of sound are, as much as possible, described in a simplified but precise manner. The fact, for example, that a particular sound is composed of harmonics (strings, pipes, etc.) or of partials (bells, timpani, etc.) finds its causes in the physical laws that govern the behavior of materials, laws that induce mathematical equations, the nature of which leads to a certain characteristic of the produced sound.

This book is intended for any reader interested in sound and music, and with a basic scientific background: students, teachers, researchers, people who work in a scientific or technical field. It describes and relies on concepts of acoustics, mathematics, psychoacoustics, computer science and signal processing, but only to the extent that this is useful in describing the subject. In order to broaden its reach, it was written in such a way that the reader may understand sound phenomena with simple analytical tools and the smallest possible amount of required knowledge. Those who teach this material will find diverse and motivating study problems, and students will find ideas for different kinds of 'projects' they may encounter in their undergraduate and graduate studies. In the end, my greatest wish would be to succeed in sharing with the reader the pleasure I find in understanding the basic mechanisms underlying the manifestation and the perception of the sound and music phenomenon.

After an introduction to acoustics, a bit of music theory, and a study of sounds and their representation in Chapter 1, we will discuss vibrational modes and the timbre of a few typical instruments in Chapter 2, and in Chapter 3, we will relate this with the question of scales and tuning systems. After wandering off into psychoacoustics in Chapter 4, and using the opportunity to discover a beautiful acoustic illusion, we will discuss several aspects of digital sound in Chapters 5 and 6: sampling, compression technology based on the properties of hearing (such as the widely known MP3 format), sound effects (vibrato, reverberation, the Leslie effect) and synthesized sounds, such as for example those produced using the Chowning technique, made popular by DX7 synthesizers.

For further development, each chapter ends with the following:

– study problems, to explore certain themes, or to study them further in depth. For the reader's information, the difficulty and the amount of work required are indicated with stars: (*) means easy, (**) is average and (***) is difficult;

– practical applications meant to be carried out on a computer, where the reader will create different kinds of sounds and play them on a crude synthesizer, experimenting on the phenomena described in the book, as well as put his or her hearing to the test, and practice his or her scales! Practical instructions relevant to these applications are given at the end of the first chapter.

Website. A website is available to illustrate the book. It contains many examples of sounds, as well as the programs used to generate them. It also contains the programs and sound files necessary to perform the practical applications, along with the answers. The address of the website is:

`www-gmm.insa-toulouse.fr/~guillaum/AM/`

Throughout the book, it will be referred to simply as the `AM` website.

Reading advice. The chapters were written in a particular, logical order, and the concept and methods developed in a given chapter are assumed to be understood in the chapters that follow. For example, the approach used to go from the wave equation to the Helmholtz equation, which is detailed in Chapter 1, will not be explained again when studying the vibrations of sonorous bodies in Chapter 2. However, you can also browse through it in any other order, referring if necessary to the previous chapters, and using the cross-references and the index to easily find where a given concept was discussed. Finally, because some phenomena are easier heard than explained, listening to the website's audio examples should shed light on any areas that may still be unclear!

Philippe GUILLAUME

Chapter 1

Sounds

Sound and air are closely related: it is common knowledge that the Moonians (the inhabitants of the Moon) have no ears! This means we will begin our study of sound with the physics of its traveling medium: air. Sounds that propagate through our atmosphere consist of a variation of the air's pressure $p(x, y, z, t)$ according to position in space x, y and z and to the time t. It is these variations in pressure that our ears can perceive. In this chapter, we will first study how these sounds propagate as waves. We will then describe a few different types of sounds and various ways of representing them. Finally, we will explain the concept of filtering, which allows certain frequencies to be singled-out.

1.1. Sound propagation

The propagation of a sound wave can occur in any direction, and depends on the obstacles in its path. We will essentially be focusing on *plane waves*, that is to say waves that only depend on one direction of space. We will assume that this direction is the x-axis, and therefore that the pressure $p(x, y, z, t)$ is independent of y and z. Hence it can simply be denoted by $p(x, t)$. This type of function represents a plane wave propagating through space, but also a sound wave inside a tube (see Figure 1.1), such as for example the one propagating through an organ pipe.

1.1.1. *A look at the physical models*

The propagation of sound through air is governed by the wave equation (see page 21), an equation we will come across several times since it also determines the movement of sound waves in the vibrating parts (strings, membranes, tubes...)

of many instruments. In the following paragraphs, we will see that, in the case of air, this equation is inferred from three fundamental equations of continuum mechanics.

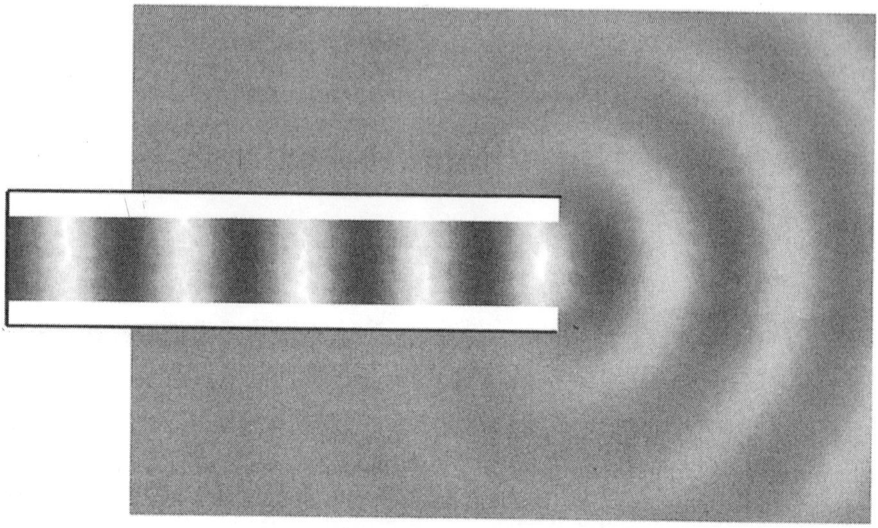

Figure 1.1. *Pressure waves in a tube open at its right end, with pressure imposed at the other end*

Along with the pressure $p(x,t)$, we rely on two other variables to describe the state of air: its density $\rho(x,t)$, and the *average* speed $v(x,t)$ of the air molecules set in motion by the sound wave, which is not to be confused with the norm of the *individual* speed of each molecule due to thermal agitation, the magnitude of which is close to that of the speed of sound, denoted by c. In the case of the plane wave that we are studying, the air moves in a direction parallel to the Ox-axis, and both the speed v, and the pressure are independent of y and z. In the absence of an atmospheric perturbation, v varies around the average value 0, and p and ρ vary around their average values p_0 and ρ_0 (see section 1.1.2), that is to say, their values in the equilibrium state: silence.

1.1.1.1. *Mass conservation*

In a *fixed* section of space, bounded by a cylinder with its axis parallel to the Ox axis and the two surfaces S_a and S_b, with respective x-coordinates a and b and areas S (see Figure 1.2), the variation of the air mass $m(t)$ is due to the amount of air going through the two surfaces. Nothing goes through the other interfaces, because the speed is parallel to the Ox axis. The air mass located inside the section is

$$m(t) = S \int_a^b \rho(x,t)\,dx,$$

and the variation of the air mass per unit of time is the derivative of $m(t)$, denoted by $m'(t)$. The incoming *flux* through S_a, that is to say, the amount of air entering the section per unit of time, is equal to $S\rho(a,t)v(a,t)$. As for the incoming flux through S_b, it is equal to $-S\rho(b,t)v(b,t)$, the change of sign being due to the fact that we are calculating the balance of what is *entering* the section (and not of what is going from left to right). The total flux is therefore

$$\Phi(t) = S[\rho(a,t)v(a,t) - \rho(b,t)v(b,t)].$$

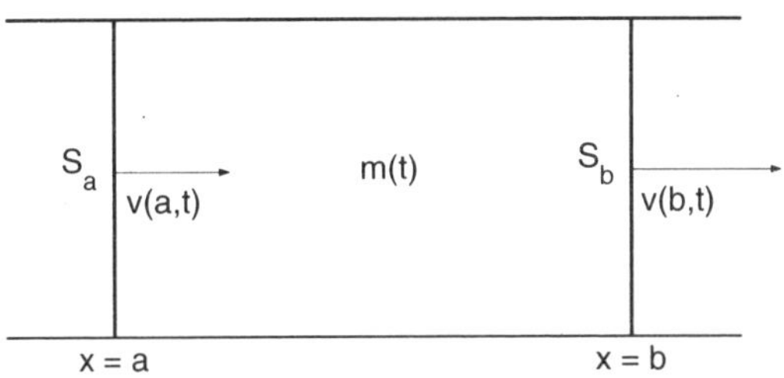

Figure 1.2. *Mass balance in the air section: there is no disappearance or creation of air!*

The fact that the total flux $\Phi(t)$ is the derivative of the mass $m(t)$,

$$\Phi(t) = m'(t),$$

can be expressed, if ∂_t denotes the partial derivative with respect to t, by

$$S[\rho(a,t)v(a,t) - \rho(b,t)v(b,t)] = S\int_a^b \partial_t\rho(x,t)\,dx.$$

If we divide by $b-a$ and if $b-a$ tends to 0 (calculation of the derivative with respect to the first argument), then after dividing both sides of the equation by x (who was on parole, confined between a and b):

$$-\partial_x(\rho(x,t)v(x,t)) = \partial_t\rho(x,t). \tag{1.1}$$

The linear acoustics hypothesis consists of assuming that the variations with respect to the equilibrium state are small, hence the use of the parameter ε, assumed to be 'small':

$$v(x,t) = \varepsilon v_1(x,t), \quad \rho(x,t) = \rho_0 + \varepsilon\rho_1(x,t).$$

If we substitute these two expressions in (1.1), and if we neglect ε^2, we get the conservation of mass equation, also called *continuity equation*:

$$\partial_t\rho_1(x,t) + \rho_0\partial_x v_1(x,t) = 0. \tag{1.2}$$

1.1.1.2. *The Euler equation*

We are now going to observe an amount of air as it moves: the section of air contained between the surfaces $S_{a(t)}$ and $S_{b(t)}$, with x-coordinates $a = a(t)$ and $b = b(t)$, respectively (see Figure 1.3), which follow the average movement of the air molecules; their derivatives are therefore such that

$$a'(t) = v(a,t) \text{ and } b'(t) = v(b,t).$$

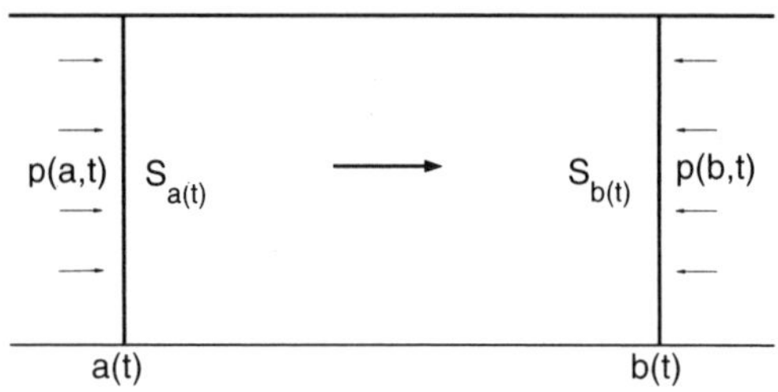

Figure 1.3. *The air section shown above is migrating. Its acceleration results from the pressure forces applied to the two surfaces $S_{a(t)}$ and $S_{b(t)}$*

The external force applied through the surface $S_{a(t)}$ to the air section is equal to $S\,p(a,t)$, and the one applied through the surface $S_{b(t)}$ is equal to $-S\,p(b,t)$. For the other interfaces, the forces cancel each other out since p is independent of y and z. We now write Newton's second law of motion $F = d(mv)/dt$:

$$S[p(a,t) - p(b,t)] = \frac{d}{dt}\left(S\int_{a(t)}^{b(t)} \rho(x,t)\,v(x,t)\,dx\right)$$

$$= S\left(\rho(b,t)\,v(b,t)\,b'(t) - \rho(a,t)\,v(a,t)\,a'(t) + \int_{a(t)}^{b(t)} \partial_t(\rho(x,t)\,v(x,t))\,dx\right)$$

$$= S\left(\rho(b,t)v^2(b,t) - \rho(a,t)v^2(a,t) + \int_{a(t)}^{b(t)} \partial_t(\rho(x,t)\,v(x,t))\,dx\right).$$

If we divide by $b - a$ and by S, and if $b - a$ tends to 0, this leads us to:

$$-\partial_x p(x,t) = \partial_x(\rho(x,t)v^2(x,t)) + \partial_t(\rho(x,t)v(x,t)).$$

If we still assume that variations with respect to the equilibrium state are small, with

$$p(x,t) = p_0 + \varepsilon p_1(x,t),$$

we get, by neglecting the ε^2 terms and those of higher order, the *Euler equation*:

$$-\partial_x p_1(x,t) = \rho_0 \partial_t v_1(x,t). \tag{1.3}$$

1.1.1.3. *The state equation*

By assuming that there are no heat transfers from one air section to the other or with the outside, or in other words that compression and expansion are *adiabatic* (a hypothesis confirmed by experiment if these effects are fast enough), the *state equation* expresses the fact that pressure variations are proportional to variations in density:

$$p_1(x,t) = c^2 \rho_1(x,t). \tag{1.4}$$

This equation also means that air has an elastic behavior: it acts like a spring. A constant c has appeared, we will see later that it represents the speed of sound. If we substitute this equation in (1.2), we find another expression for the state equation:

$$\partial_t p_1(x,t) + c^2 \rho_0 \partial_x v_1(x,t) = 0. \tag{1.5}$$

1.1.2. *The wave equation*

We now have at our disposal all the tools necessary to describe the movement of sound waves through air. If we differentiate the state equation (1.5) with respect to time and the Euler equation (1.3) with respect to x, we get

$$\partial_{t^2} p_1(x,t) = -c^2 \rho_0 \partial_{tx} v_1(x,t),$$
$$\partial_{x^2} p_1(x,t) = -\rho_0 \partial_{tx} v_1(x,t).$$

The expression ∂_{t^2} indicates two differentiations with respect to time, ∂_{tx} indicates one differentiation with respect to time and another with respect to x, and so on. All we have to do now is compare these two equations to obtain the *wave equation*:

$$\partial_{t^2} p_1(x,t) = c^2 \partial_{x^2} p_1(x,t). \tag{1.6}$$

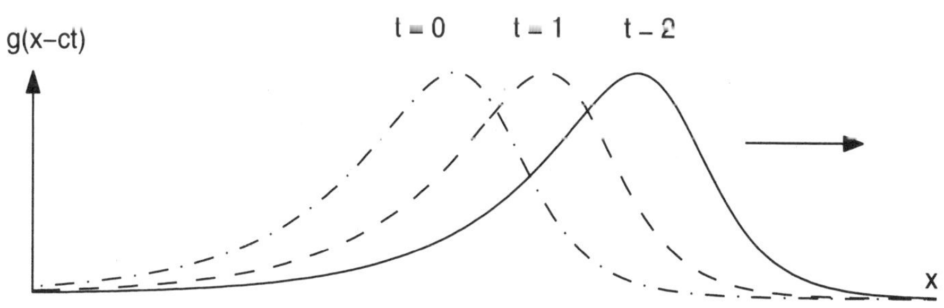

Figure 1.4. *Three snapshots of a traveling plane wave along an axis*

A mathematical analysis of this equation shows that the general solution is of the form

$$p_1(x,t) = g(x - ct) + h(x + ct).$$

The function $g(x - ct)$ keeps a constant value in the case of a point in motion, the trajectory of which is such that $x - ct$ = constant (such a trajectory is called a *characteristic trajectory*); thus $g(x - ct)$ represents a *traveling* wave propagating along the x-axis at the *speed of sound* c from left to right (Figure 1.4 shows the usual orientation for the axis). Likewise, the function $h(x + ct)$ is constant at the points with x-coordinates such that $x + ct$ = constant, and in that case represents a traveling wave propagating at the speed c from right to left. For air at a temperature T expressed in Kelvin (with 32°F = 0°C = 273 K), the approximate values for the speed of sound, the density and the atmospheric pressure (in *pascals* and in *bars*) are

$$c = 20\sqrt{T}, \quad \rho_0 = \frac{353}{T}, \quad p_0 = 1.013\,10^5 \text{ Pa} = 1.013 \text{ bar at } 0°\text{C},$$

$$c = 330 \text{ m/s at } 0°\text{C}, c = 340 \text{ m/s at } 16°\text{C}.$$

For example, the functions

$$u_+(x,t) = \sin(kx - 2\pi ft),$$
$$u_-(x,t) = \sin(kx + 2\pi ft),$$

with $k = 2\pi f/c$, are solutions to the wave equation. They are periodic with respect to variables of time and space. The space period

$$\lambda = \frac{2\pi}{k} = \frac{c}{f}$$

is called the *wavelength*. It is one of the most elementary forms of musical sound, with a pitch, or a *frequency* f, measured in *hertz* ($1\,\text{Hz} = 1\,\text{s}^{-1}$), a unit named after physicist H. R. Hertz, and with a *timbre* (the sound's 'color') similar to that of a recorder (a type of flute).

These two functions u_+ and u_- propagate in opposite directions. Adding the two leads to an interesting wave, also a solution to the wave equation:

$$p_1(x,t) = \sin(kx - 2\pi ft) + \sin(kx + 2\pi ft)$$
$$= 2\sin(kx)\cos(2\pi ft).$$

As you can see, for all points $x = n\pi/k$, $n \in \mathbb{Z}$ (the set of integers), for which $\sin(kx) = 0$, the pressure $p = p_0 + \varepsilon p_1$ is constant and equal to p_0: these points are called vibration *nodes*, whereas for points $x = (n + 1/2)\pi/k$, $n \in \mathbb{Z}$, the pressure $p(x,t) = p_0 \pm 2\varepsilon\cos(2\pi ft)$ undergoes its maximum amplitude variations: these points are called *antinodes*. Such waves are referred to as standing waves (see Figure 1.5).

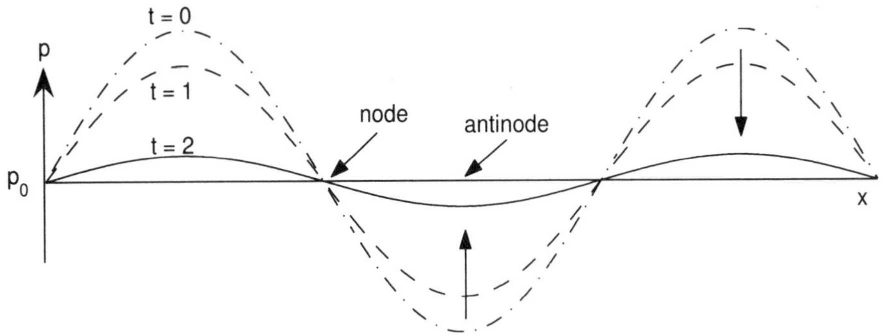

Figure 1.5. *Three snapshots of a standing plane wave*

1.1.3. *The Helmholtz equation*

In physics, a wave containing only one frequency, *i.e.* of the form

$$p_1(x,t) = \varphi(x)\exp(2i\pi ft)$$

where φ can also be a complex function[1] and where $f \in \mathbb{R}$ (set of real numbers), is said to be harmonic. The real and imaginary parts of such a wave are also harmonic. Functions of the form

$$p_1(x,t) = \varphi(x)\psi(t) \tag{1.7}$$

are said to be *separated variable* functions. Additionally, if φ is real, the wave is referred to as a *standing* wave: except for a real multiplicative factor $\varphi(x)$, all points simultaneously undergo the same variation in pressure $\psi(t)$.

If we substitute equation (1.7) in (1.6), we get, after dividing by $\varphi(x)\psi(t)$,

$$\frac{\psi''(t)}{\psi(t)} = c^2\frac{\varphi''(x)}{\varphi(x)}.$$

1. The use of complex numbers and functions makes the notations simpler. The physical signal associated with a complex function can be obtained simply by calculating the real part of that function. The sign of f indicates whether $p_1(x,t)$ travels clockwise or counterclockwise along the unit circle's circumference. When switching over to real numbers, because $\cos(-x) = \cos(x)$ and $\sin(-x) = -\sin(x)$, the frequency can always be assumed to be ≥ 0, by changing the sign of the sine term if necessary.

This expression cannot vary, since the term on the left depends only on time, and the one on the right depends only on x. Hence it is a constant, which will be denoted by $-(2\pi f)^2$, where f is an arbitrary real number[2]. Thus, on the one hand, we get

$$\psi''(t) + (2\pi f)^2\psi(t) = 0,$$

the general solution of which is

$$\psi(t) = A\exp(2i\pi ft) + B\exp(-2i\pi ft).$$

If $B = 0$ or $A = 0$, the wave is harmonic with frequency $\pm f$. On the other hand, if we define $k = 2\pi f/c$, called the *wavenumber*, we obtain the homogenous *Helmholtz equation*:

$$\varphi''(x) + k^2\varphi(x) = 0, \tag{1.8}$$

the general solution of which is

$$\varphi(x) = \alpha\exp(ikx) + \beta\exp(-ikx).$$

Thus, the harmonic pressure waves with frequency f are of the form

$$p_1(x,t) = [\alpha\exp(ikx) + \beta\exp(-ikx)]\exp(2i\pi ft),$$

where the constants α, $\beta \in \mathbb{C}$ (set of complex numbers) are determined by the conditions imposed at the interfaces with objects. As for standing harmonic waves with frequency f, they are of the form

$$p_1(x,t) = \alpha\sin(k(x - x_0))\exp(2i\pi ft),$$

where x_0 is one of the vibration nodes.

If we follow the same process (*i.e.* start with three fundamental equations), we come to the conclusion that, in the general case, when the waves are not necessarily plane waves, the pressure is a solution to the three dimensional wave equation

$$\partial_{t^2}p_1(x,y,z,t) = c^2\Delta p_1(x,y,z,t) \tag{1.9}$$

where $\Delta = \partial_{x^2} + \partial_{y^2} + \partial_{z^2}$ is called the *Laplacian*, and the Helmholtz equation becomes

$$\Delta\varphi(x,y,z) + k^2\varphi(x,y,z) = 0.$$

2. Choosing a negative constant allows us to pick only the physical solutions we are interested in: functions that are sinusoidal when plotted versus time, *i.e.* the harmonic ones. Others exist, but we will not be using them.

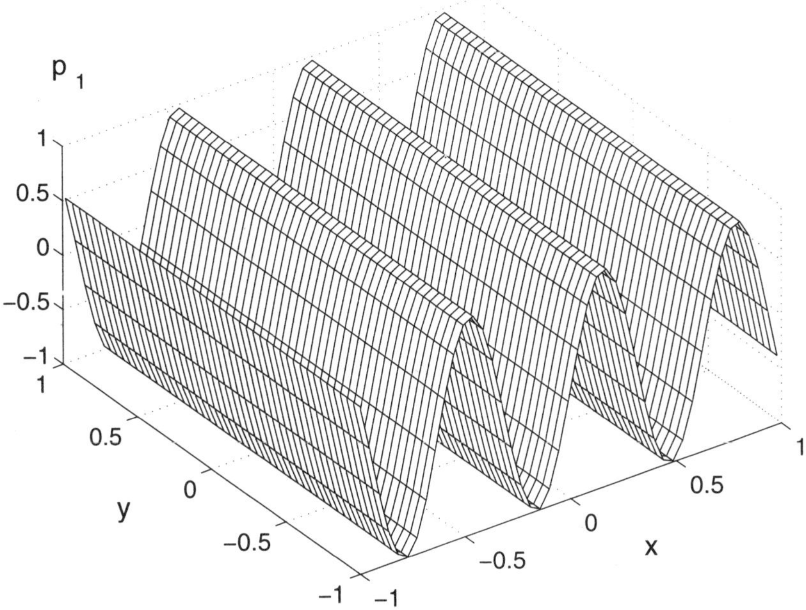

Figure 1.6. *A harmonic plane wave. It propagates along the Ox-axis (2D section) without any damping*

For example, *spherical harmonic waves*, produced by a punctual source assumed to be placed at the origin, are of the type (with $r = \sqrt{x^2 + y^2 + z^2}$):

$$p_1(x, y, z, t) = \alpha \frac{\exp(ikr - 2i\pi ft)}{r}.$$

These waves are called spherical because, for a set value of t, given a sphere with its center at the origin, the pressure is the same at every point on the sphere. Note that these are not standing waves.

1.1.4. *Sound intensity*

Earlier in this chapter, we denoted the pressure (in the case of a plane wave) by $p(x, t) = p_0 + \varepsilon p_1(x, t)$ where p_0 is the pressure in the equilibrium state, or average pressure. The difference $p(x, t) - p_0$ is called the *acoustic pressure* p_a:

$$p_a(x, t) = p(x, t) - p_0.$$

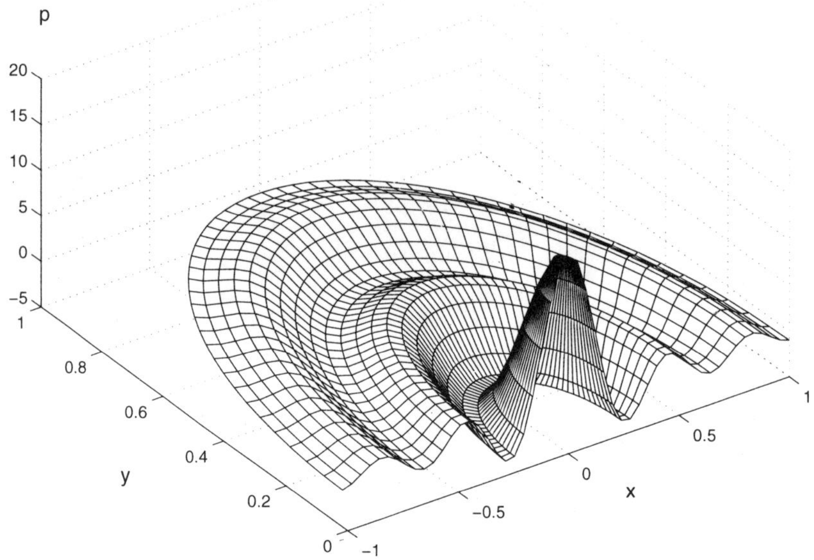

Figure 1.7. *A spherical harmonic wave (2D section). It decreases as $1/r$*

In order to set air in motion, a certain amount of energy had to be provided. The propagation of the air deformation corresponds to the propagation of the initial energy. The term *sound intensity* (or acoustic intensity) refers to the mean power – with respect to time – of the acoustic wave per unit of surface. It is measured in watts/m^2, depends on the point where it is measured, x, and is obtained from the formula

$$I(x) = \frac{1}{T}\int_0^T p_a(x,t)v(x,t)\,dt$$

where the time scale T depends on the context. This integral can be equal to zero if for example p_a and v are in quadrature (a difference in phase equal to $\pi/2$). In the case of a traveling plane wave $p_a(x,t) = g(x - ct)$, the calculations based on the Euler equation (1.3) and the state equation (1.5) lead to $v(x,t) = p_a(x,t)/c\rho_0$, hence

$$I(x) = \frac{1}{Tc\rho_0}\int_0^T p_a^2(x,t)\,dt. \tag{1.10}$$

In the case of a harmonic wave, for example $p_a(x,t) = p_\alpha \sin(kx - 2\pi ft)$, we get, for $T = 1/f$,

$$I(x) = \frac{p_\alpha^2}{2c\rho_0} \simeq \frac{p_{\text{eff}}^2}{415},$$

with $p_{\text{eff}} = p_\alpha/\sqrt{2}$, a formula often used when calculating the intensity. In the case of a spherical wave $p_a(x,y,z,t) = p_\alpha \sin(kr - 2\pi ft)/r$, the calculation for a high enough value of r leads to

$$I(x,y,z) \simeq \frac{p_\alpha^2}{2c\rho_0} \times \frac{1}{r^2}.$$

Therefore the intensity of a sound originating from a punctual source (in the absence of damping) is inversely proportional to the square of the distance from that source.

The *hearing threshold* is approximately

$$I_0 = 10^{-12}\ \text{W/m}^2,$$

the normal level for a conversation is $1.2\,10^{-5}$ W/m^2 and the *pain threshold* is about 1 W/m^2. We will see in Chapter 4 that these threshold values depend on certain parameters, particularly frequency. Notice the impressive value for the ear's dynamic range, 10^{12}! Rather than W/m^2, the preferred unit is the bel (named after Alexander Graham Bell, a professor at a school for the hearing impaired, and the inventor of the telephone) or the *decibel*, a dimensionless unit that measures the tenth of the base 10 logarithm of a ratio to a given threshold, the hearing threshold for example. If the sound intensity is denoted by L_I, then this can be expressed as:

$$L_I = 10 \log \frac{|I|}{I_0}\ \text{dB}.$$

As a consequence, the hearing threshold is set at 0 dB, the pain threshold is 120 dB, and for a conversation, it is equal to 70 dB. Note that at some rock concerts, the intensity sometimes exceeds 140 dB!

Small question: what happens to a symphonic orchestra when the number of violins is multiplied by 10?

Answer (see section 1.6.3): a 10 dB increase in the sound level. In other words, going from 1 to 10 violins leads to the same volume increase as when going from 10 to 100 violins! This is an example of Fechner's law, named after the German physiologist Gustav Fechner: *a sensation varies proportionally to the logarithm of the stimulus* (see [LEI 80], but also [ZWI 81] for a more moderate point of view, discussed in Chapter 4).

1.2. Music theory interlude

Before we go any further, it may be necessary to brush up on a few elementary concepts of music theory and the relevant vocabulary. A musical note is characterized by three main parameters: its pitch, its duration and its intensity. Here we will be focusing on the pitch. The pitch is directly related to the note's frequency[3]: low frequencies correspond to *low-pitched* sounds, and high frequencies to *high-pitched* sounds. The reference frequency for a musician is the A at 440 Hz, the note made by a *tuning fork*, and also the note used for most dial tones.

1.2.1. *Intervals, octave*

In music theory, the distance between two different notes is called an *interval.* When our ears estimate the interval between two notes, what affects their perception is the *ratio* of their frequencies, and not the *difference* in frequencies. This is another example of Fechner's law, previously mentioned in regards to intensity: our sensation of pitch varies proportionally to the logarithm of the frequency (this law does not apply to extremely high-pitched and low-pitched sounds, as we will see in Chapter 4). For example, the two musical intervals $[110 \text{ Hz}, 220 \text{ Hz}]$ and $[220 \text{ Hz}, 440 \text{ Hz}]$ are perceived as *equal* because the frequency ratios are equal: $220/110 = 440/220 = 2$, whereas in the mathematical sense, the second interval is twice as long as the first: $440 - 220 = 2 \times (220 - 110)$. The interval between two notes, in the case where their frequency ratio is equal to two, is called an *octave*.

1.2.2. *Scientific pitch notation*

The sounds produced by two notes one octave apart from each other are very similar (we will see why in section 2.1.2), to the point that they are referred to as the same note. The frequency 880 Hz, for example, one octave above the A of a tuning fork, will also produce an A, but at a higher pitch. In order to tell them apart, we will use the scientific pitch notation: the 440 Hz A will be denoted by A4, the next one at 880 Hz will be denoted by A5, followed by A6 at $1,760$ Hz. Likewise, in the other direction we have A3 at 220 Hz, then A2 at 110 Hz, and so on. The same goes for other notes, number 4 being used for the notes found between the C at 261.6 Hz and the B at 493.9 Hz, all of which are located near the middle of a piano keyboard.

1.2.3. *Dividing the octave into twelve semitones*

Other intervals are determined by the choice of a *tuning system*, which precisely sets the frequency ratios of the notes to one another, and will be studied in detail in Chapter 3. Here, we will be considering the case of *equal temperament*. For this tuning system, the octave is divided into twelve equal intervals called *semitones*, with frequency ratios of $2^{1/12} \simeq 1.0595$. If we start at a note with frequency f, and go up a semitone twelve times, we get, one after the other, the notes with frequencies $2^{1/12} f$,

3. As we will see in the next section, the pitch is the fundamental frequency of a note.

then $2^{1/12} \times 2^{1/12}f = 2^{2/12}f$, then $2^{3/12}f$, ..., $2^{11/12}f$, and finally $2^{12/12}f = 2f$, bringing us, as expected, to the octave above by equal intervals.

These thirteen notes comprise what is called the *chromatic scale*, invented by the Chinese over 4,000 years ago! Starting with C, these notes are C, C♯, D, D♯, E, F, F♯, G, G♯, A, A♯, B, C, the ♯ sign indicating that the note has been raised by a semitone; the resulting note is said to be *altered*. If we use the ♭ to lower the note by a semitone, this series of notes can also be written C, D♭, D, E♭, E, F, G♭, G, A♭, A, B♭, B, C. The notes C♯ and D♭ are said to be *enharmonics* and are equal in equal temperament. The same goes for the other enharmonic notes, D♯ and E♭, F♯ and G♭, etc. Note however that musicians who have the possibility of determining themselves the pitch of a note, such as violinists, often play the C♯ slightly above the D♭. The combined interval of two semitones is, of course, called a tone, and there are six tones in an octave. The corresponding notes comprise the *whole tone scale*, abundantly used by Claude Debussy.

1.2.4. *Diatonic scales*

The usual scales are neither the chromatic scale nor the whole tone scale, but instead the *diatonic* scales, based on both types of intervals: the tone and the semitone, and comprising eight notes, the last of which is one octave higher than the first. These scales are the result of placing two *tetrachords* (four consecutive notes) one after the other, where each tetrachord must include two tones, hence the name diatonic. Where the semitone is placed inside each tetrachord then determines the different possible modes or scales. The notes of a scale are called the *degrees* of the scale. The first (and the eighth, since it is the 'same' note) is called the *tonic*, the fifth is the *dominant* and the seventh is the *leading tone*, the degree that 'leads' to the tonic in tonal harmony.

There are several types of diatonic scales, two of which have played a central role in all of classical music: the *major* scale and the *minor* scale. The essential differences between these two scales are their third and sixth degrees, called *tonal* notes for this reason. Bright, upbeat or joyful themes (marches, festive themes, dances) are often written in the major scale, whereas mournful, sad or gloomy themes (requiems, nocturnes funeral marches) are usually written in the minor scale. The other scales are called *modal* scales, and were widely used all through the Middle Ages, particularly in ecclesiastical music.

1.2.4.1. *Major scale*

A diatonic major scale is comprised of the following intervals: tone, tone, semitone, tone, tone, tone, semitone. Starting with C for example, we get the sequence of notes C, D, E, F, G, A, B, represented on the staff as follows (see Figure 1.8).

Figure 1.8. *C major scale, beginning with C4 and ending with B4*

We owe this notation process to Guy d'Arezzo (early 11th century). The scale's different degrees are placed, in turn, *on* and *between* the staff's lines. The intervals C–D, C–E, C–F, ... , C–B are called the *second, major third, fourth, fifth, major sixth* and *major seventh*, respectively. These names refer of course to the intervals between the notes and not to the notes themselves. Hence the intervals F–A, G–B also form a major third, comprising two tones, and the intervals D–A, E–B and F–C form a fifth, comprised of three and a half tones and corresponding to the frequency ratio $2^{7/12} \simeq 1.5$.

1.2.4.2. *Minor scales*

There are two types of minor diatonic scales, depending on whether the melody ascends or descends:

– the *ascending melodic* minor scale, comprising the intervals tone, semitone, tone, tone, tone, tone, semitone. Starting with C, we get the sequence of notes C, D, E♭, F, G, A, B, C;

– the *descending melodic* minor scale, comprising the intervals tone, semitone, tone, tone, semitone, tone, tone. If we choose C as the tonic, we get the notes C, D, E♭, F, G, A♭, B♭, C. If we choose A as the tonic, this leads to the notes A, B, C, D, E, F, G, A, a scale without any alterations. This last scale is called the *relative* minor scale of the C major scale.

Figure 1.9. *Ascending melodic C minor scale*

Figure 1.10. *Descending melodic C minor scale*

The somewhat different *harmonic* minor scale is used, hence the name, to compose the chords (the harmony) meant as the accompaniment to a melody composed in a minor scale. It is comprised of the intervals tone, semi-tone, tone, tone, semi-tone, one tone and a half, semitone. Starting with C, we get the sequence of notes C, D, E♭, F, G, A♭, B, C.

Figure 1.11. *Harmonic C minor scale*

The intervals C–E♭, C–A♭ and C–B♭ are called the *third minor*, *sixth minor* and *seventh minor*, respectively, and represent intervals of one tone and a half, four tones and five tones.

1.3. Different types of sounds

A listener located at a given point in space can perceive the variation in pressure at that point. This variation is a function of time, and will be denoted by $s(t) = p_a(x,t)$. It is the *sound signal*. The sinusoidal signals discussed earlier can be expressed differently depending on the context:

$$\begin{aligned} s(t) &= \alpha \cos(\omega t + \theta) \\ &= \alpha \cos(2\pi f t + \theta) \\ &= \alpha \operatorname{Re}(\exp(i\theta)\exp(2i\pi f t)) \\ &= a\cos(2\pi f t) + b \sin(2\pi f t) \\ &= c_1 \exp(2i\pi f t) + c_2 \exp(-2i\pi f t) \end{aligned}$$

where $\operatorname{Re}(z)$ refers to the real part of z, and:

– $\omega \geq 0$ is the *pulsation* in radians/s;

– $f = \omega/2\pi \geq 0$ is the *frequency*[4] expressed in *hertz* (Hz); it indicates the number of vibrations per seconds;

– $\alpha \geq 0$ is the *amplitude*;

– θ is the *phase* at $t = 0$, measured in radians with $\theta \in [0, 2\pi[$;

– $a = \alpha \cos\theta$ (choose $t = 0$), $b = \alpha\cos(\theta + \pi/2)$ (choose $2\pi f t = \pi/2$);

– $c_1 = (a - ib)/2$, $c_2 = (a + ib)/2 = \overline{c_1}$ (use $\exp(ix) = \cos x + i \sin x$).

This sinusoidal sound is one of the simplest possible sounds, it is said to be a *pure sound*. For a plane wave of the form $p_a(x,t) = \alpha\cos(kx + 2\pi f t)$, it is the sound produced at all points x such that $kx = \theta + 2n\pi$, $n \in \mathbb{Z}$.

---

4. In the case of a real function, the frequency is always assumed to be positive or equal to zero. For complex functions, because $\cos(2\pi f t) = [\exp(2\pi i f t) + \exp(-2\pi i f t)]/2$, we also have to consider negative frequencies (in this case $-f$).

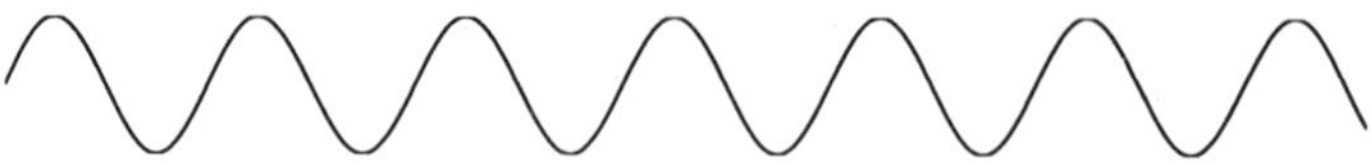

Figure 1.12. *A sinusoidal sound, considered a 'pure sound'*

One of the important properties of the wave equation is that it is linear and homogenous. As a result, if $p_1(x,t) = \alpha_1 \cos(k_1 x + 2\pi f_1 t)$ and $p_2(x,t) = \alpha_2 \cos(k_2 x + 2\pi f_2 t)$ are solutions to this equation (which is the case for $k_1 = 2\pi f_1/c$ and $k_2 = 2\pi f_2/c$), then $p_1(x,t) + p_2(x,t)$ will also be a solution to the wave equation. At a given point x, the perceived sound will then be of the form $s(t) = \alpha_1 \cos(2\pi f_1 t + \theta_1) + \alpha_2 \cos(2\pi f_2 t + \theta_2)$. This more complex sound is the result of superposing the two frequencies f_1 and f_2. Taking this process further shows that an acoustic wave can, at a certain point in space, produce a sound signal of the form

$$\begin{aligned} s(t) &= \alpha_1 \cos(2\pi f_1 t + \theta_1) + \alpha_2 \cos(2\pi f_2 t + \theta_2) + \dots \\ &= \sum_{n\geq 1} \alpha_n \cos(2\pi f_n t + \theta_n). \end{aligned} \tag{1.11}$$

If the sum is comprised of an infinite number of terms, it can only be evaluated if the α_n and the f_n meet certain convergence conditions. The *spectrum* of such a sound, that is to say, the set of frequencies f_n contained in this sum, is said to be a *discrete spectrum*.

A young person is usually considered to be able to perceive frequencies from 20 Hz to 20 kHz, and sounds become inaudible outside this interval (infrasounds or ultrasounds).

1.3.1. *Periodic sounds*

Let us again consider sounds of the form (1.11). An interesting case occurs when all of the frequencies are *integer* multiples of a given frequency $f > 0$: $f_n = nf$. In such a case, the sound signal

$$s(t) = \sum_{n\geq 1} \alpha_n \cos(2\pi n f t + \theta_n)$$

is *periodic* with *period* $T = 1/f$, that is to say that $s(t+T) = s(t)$ for any t. This is because

$$\cos(2\pi n f(t+T) + \theta_n) = \cos(2\pi n f t + 2n\pi + \theta_n) = \cos(2\pi n f t + \theta_n).$$

There are, of course, no truly periodic sounds, if only because no sounds could have begun before the Big Bang!

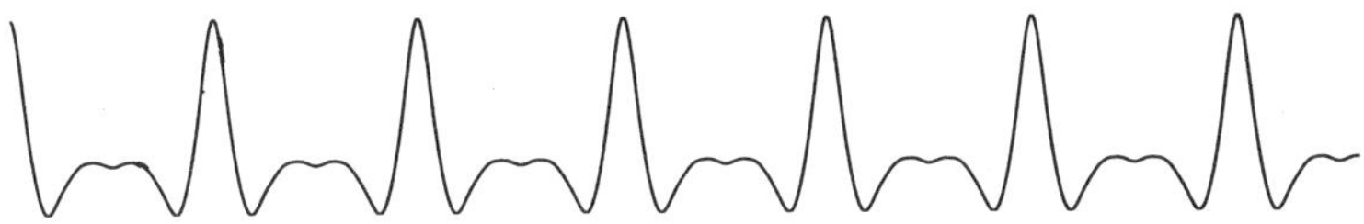

Figure 1.13. *Periodic sound (approximation of the sound of a trumpet)*

In music, the frequency f (and likewise the component $\cos(2\pi ft + \theta)$) is called the *fundamental* – hence it determines the pitch of the corresponding note – and the frequency $f_n = nf$ is called the *n-th harmonic* (not to be confused with a *harmonic* wave). Therefore, the first harmonic is also the fundamental. If for example f is the frequency of C4 (261.6 Hz), then f_2 is one octave above (C5), f_3 is one fifth higher (G5), f_5 is slightly below the major third (E6), etc. (see also Figure 3.1).

$f = 261.6$ Hz	$2f$	$3f$	$4f$	$5f$	$6f$	$7f$	$8f$	$9f$
C4	C5	G5	C6	E6	G6	B♭6	C7	D7

(1.12)

A sound comprised of a high number of harmonics is perceived as 'rich' (such as for the harpsichord or the violin), whereas a sound comprised of only few harmonics will be perceived as 'poor' (such as for the recorder).

Figure 1.14 shows the signal obtained by adding all of the terms $(\sin 2\pi nt)/n$, $n = 1, 2, 3, 4$. You can listen to these sounds on the AM website. We are getting closer and closer to a *triangular* signal, the simplest model for the sound of a violin.

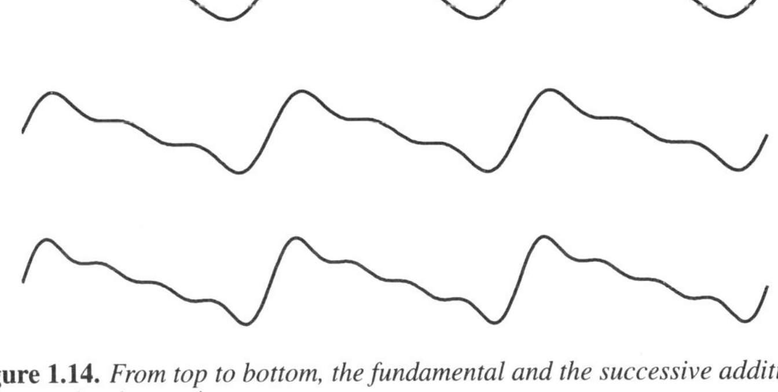

Figure 1.14. *From top to bottom, the fundamental and the successive additions of the 2nd, 3rd and 4th harmonics (simplified model for the sound of a violin)*

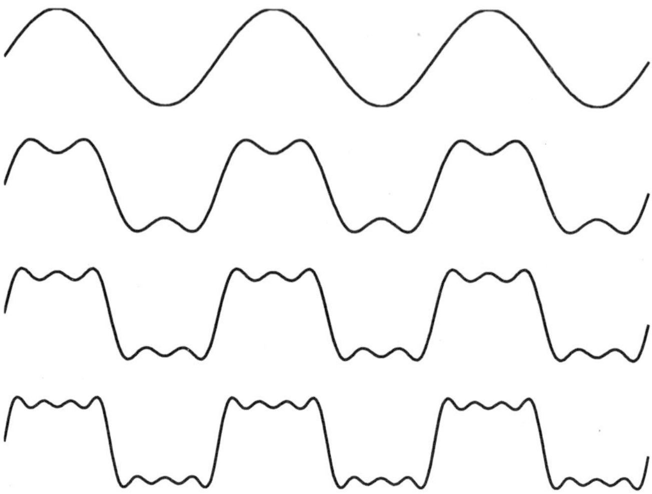

Figure 1.15. *From top to bottom, the fundamental and the successive additions of the 3rd, 5th and 7th harmonics (simplified model for the sound of a clarinet)*

Figure 1.15 (see also the AM website) was obtained in the same way, with the odd harmonics $n = 1, 3, 5, 7$. The result is closer to a *rectangular* signal, the simplest model for the sound of a clarinet.

Usually, *sustained sound* instruments, such as the violin or the organ, produce periodic sounds, at least over a significant period of time. We will now describe the mathematical tool that can allow us to analyze such sounds.

1.3.1.1. *Fourier series*

Mathematician Joseph Fourier (1768-1830) was the first to analyze periodic sounds and to decompose them into the trigonometric series that bear his name: the Fourier series. He developed this theory while he was working on heat propagation in solids.

If $s(t)$ is a T-periodic function (*i.e.* with period T), integrable[5] over the interval $[0, T]$, the *Fourier coefficients* c_n, with n an integer, are defined by

$$c_n = \frac{1}{T} \int_0^T s(t) \exp(-2i\pi nt/T)\, dt. \tag{1.13}$$

5. A function s is said to be integrable over the interval $[a, b]$ if $\int_a^b |s(t)|\, dt < \infty$.

This constitutes the Fourier *analysis*. It can be shown that, if certain additional conditions are met, the series below, called the *Fourier series*, leads back to the values of $s(t)$:

$$\sum_{n=-\infty}^{+\infty} c_n \exp(2i\pi nt/T) = s(t).$$

This constitutes the Fourier *synthesis*: the sound $s(t)$ is made up of the sum of its harmonics $c_n \exp(2i\pi nt/T)$ with frequency n/T. This sum can also be expressed with sines and cosines:

$$s(t) = \frac{a_0}{2} + \sum_{n=1}^{+\infty} a_n \cos(2\pi nt/T) + b_n \sin(2\pi nt/T)$$

with $a_n = c_n + c_{-n}$ and $b_n = i(c_n - c_{-n})$.

When $s^2(t)$ is integrable over the interval $[0, T]$, the sound's intensity (1.10) over a period is related to the Fourier coefficients by way of Parseval's theorem:

$$\frac{1}{T}\int_0^T |s(t)|^2\, dt = \sum_{n=-\infty}^{+\infty} |c_n|^2.$$

1.3.2. *Sounds with partials*

Other instruments, such as bells, most percussion instruments, as well as the piano, to a small extent, produce sounds of the form (1.11), but that are not periodic. In terms of frequency, this means that there is no frequency f such that all frequencies f_n are integer multiples of f. The Fourier series analysis can then no longer directly be applied.

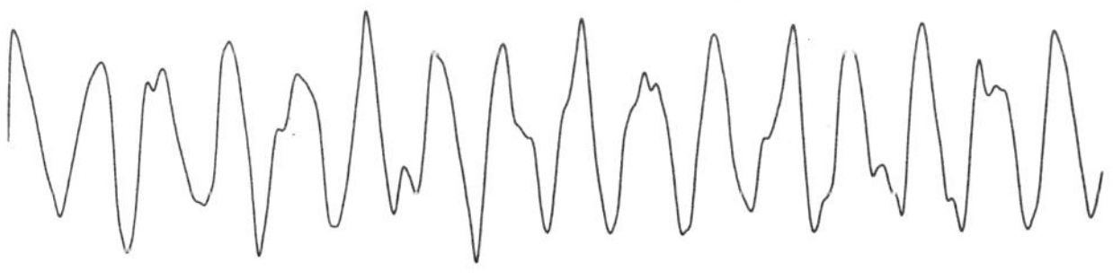

Figure 1.16. *An example of a non-periodic sound (timpani)*

The frequencies found in such a sound are called *partials*. An example of the sound of a bell, found in [PIE 99], is comprised of the frequencies[6] $0.5f_p$, f_p, $1.2f_p$, $1.5f_p$,

6. Strictly speaking, this sound is also periodic, with period $10/f_p$, since the 7 frequencies are integer multiples of $f_p/10$. But with such a consideration, the first 4 harmonics are equal to zero. Furthermore, if higher partials are added, the value $f_p/10$ will no longer be suitable. Finally, these frequencies are only approximations, it may very well be that the exact values have no common divisor.

$2f_p$, $2.5f_p$ and $3f_p$. The second partial is called the *principal*, and determines the pitch of the note. Bellfounders can tune the first 5 partials of a bell's sound by changing its thickness, and arrange them so that the third partial is a minor third (the ratio is $2^{3/12} \simeq 1.189$ in equal temperament, see Chapter 3) above the principal, which gives them their characteristic sound. The following ratios are a fifth, an octave, a major third, etc. The previous example actually does not include enough partials to produce a realistic sound. The analysis of a large bell, for $f_p = 233$ Hz (B♭3), yielded the following frequencies and their respective intensities:

$$f = [0.5,\ 1,\ 1.188,\ 1.530,\ 2.000,\ 2.470,\ 2.607,\ 2.650,\ 2.991,\ 3.367,\ 4.137,\ 4.487,\\ 4.829,\ 5.385,\ 5.863,\ 6.709,\ 8.077,\ 8.547,\ 9.017,\ 9.530,\ 11.026,\ 12.393] \times f_p,$$

$$I = [350,\ 950,\ 500,\ 150,\ 700,\ 100,\ 250,\ 370,\ 1,000,\ 180,\ 300,\ 100,\\ 150,\ 300,\ 100,\ 100,\ 50,\ 20,\ 10,\ 35,\ 5,\ 15].$$

Combined with an adequate envelope (see section 2.5.2), this information allows a better sound reconstruction, as you can hear from the examples on the AM website.

1.3.3. *Continuous spectrum sounds*

Singing is still quite close to a periodic sound, but with more complex sounds such as spoken voice, not only are we much further off from periodic sounds, we are not even dealing with a sum of *punctual* frequencies anymore, as we did in (1.11). It is in fact the opposite, a *continuous* set of frequencies. Instead of being written as a *sum* of $\exp(2i\pi f_n t)$ terms with frequency f_n, such sounds $s(t)$ can be represented using integrals of similar terms, constituting the Fourier synthesis:

$$s(t) = \int_{-\infty}^{+\infty} \widehat{s}(f) \exp(2i\pi ft)\, df, \tag{1.14}$$

where the function $\widehat{s}(f)$, called the *Fourier transform* of s, is given by the analysis formula

$$\widehat{s}(f) = \int_{-\infty}^{+\infty} s(t) \exp(-2i\pi ft)\, dt. \tag{1.15}$$

The *modulus* $|\widehat{s}(f)|$ gives the amplitude of the frequency component $\exp(2i\pi ft)$, and the argument of $\widehat{s}(f)$ gives the phase at the origin of that component[7]. The spectrum of such a sound, as opposed to the discrete spectrum in (1.11), is called a *continuous spectrum*. In particular, this representation applies to any function $s(t)$ integrable over $\mathbb{R}$ such that $\widehat{s}(f)$ is also integrable over $\mathbb{R}$.

7. Any complex number z can be expressed as $z = \rho\exp(i\theta)$ where $\rho \geq 0$ is called the *modulus* of z and $\theta \in \mathbb{R}$ is called the *argument* of z.

What does the Fourier transform have to do with the Fourier series expansion discussed in the previous section? The answer is very simple in the following case: let $s(t)$ be a sound, *equal to zero* outside an interval $[0, T]$, and consider its T-periodic extension $\widetilde{s}(t)$ defined by $\widetilde{s}(t + kT) = s(t)$ for any t belonging to $[0, T]$ and any integer k. Then, if the Fourier coefficients of $\widetilde{s}(t)$, are denoted by c_n, we have

$$c_n = \frac{1}{T}\widehat{s}\left(\frac{n}{T}\right).$$

This result can be inferred simply by applying the two corresponding definitions (1.13) and (1.15). In other words, the Fourier coefficients of $\widetilde{s}(t)$ are exactly the values, divided by T, of the Fourier transform of $s(t)$ calculated at the discrete frequencies n/T. The higher the value of T, the closer together these frequencies will be. This means that these Fourier coefficients provide valuable information about the Fourier transform itself, that is to say, the frequency content of the sound $s(t)$ (we will discuss this further in section 1.4.1). In the case where $s(t)$ has non-zero values outside the interval $[0, T]$, the c_n can be used as an approximation of the Fourier transform of $s(t)$. The error then increases as the values outside $[0, T]$ move farther away from 0.

However, this representation, solely based on frequency is not wholly satisfactory (no more in fact than (1.11)), because the function $\widehat{s}(f)$ does not provide any direct information about the sound signal's time behavior. It would be better to use a representation that involves both time and frequency, such as the one used by musicians when they transcribe a musical piece to a score. For example, Figures 1.17 and 1.18 show an analysis of Maria Callas singing in *Norma* by Bellini. We will explain in the following section how such a representation is achieved, where high intensities are displayed in darker tones and weak intensities with lighter tones. Notice the famous *vibrato* that earned her so many admirers, and how remarkably regular it is!

Finally, humans are not the only ones able to sing, here is, for those nature lovers out there, the song of a goldfinch, a bird that can be found in the French countryside, in both time and time-frequency representations in Figure 1.19...

Or, for those who prefer altitude, the song of a mountain whistler shown in Figure 1.20. Notice, incidentally, the way the time signal is 'enveloped', we will mention this again when we discuss instruments in section 2.5.2.

1.3.4. *Noise*

The nature segment is now over, we go back to 'civilization'. What surrounds us on a permanent basis? Noise! Noise is usually associated with sounds that lack any structure and are unpredictable. For example, a sound signal s where at any time t, the value of $s(t)$ is random and does not depend on values at other moments, is a noise

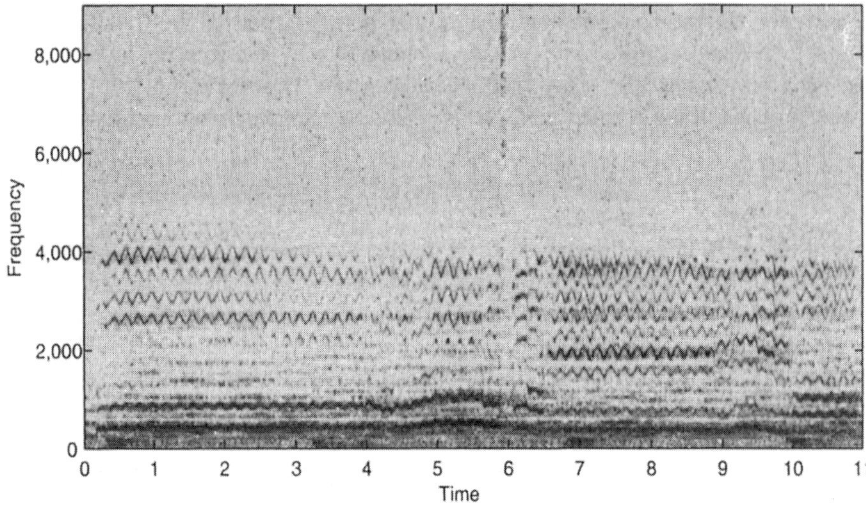

Figure 1.17. *The beginning of* Norma *by Maria Callas. The first note is an A4 (441 Hz). The first two harmonics appear very clearly. Then there is a 'gap' between* $1,000$ *and* $2,500$ *Hz, and the frequency energy reappears between* $2,500$ *and* $4,000$ *Hz, where the voice is particularly effective at being heard over the orchestra. Notice of course the vibrato, but also the 's' at* $t = 6$ *s.*

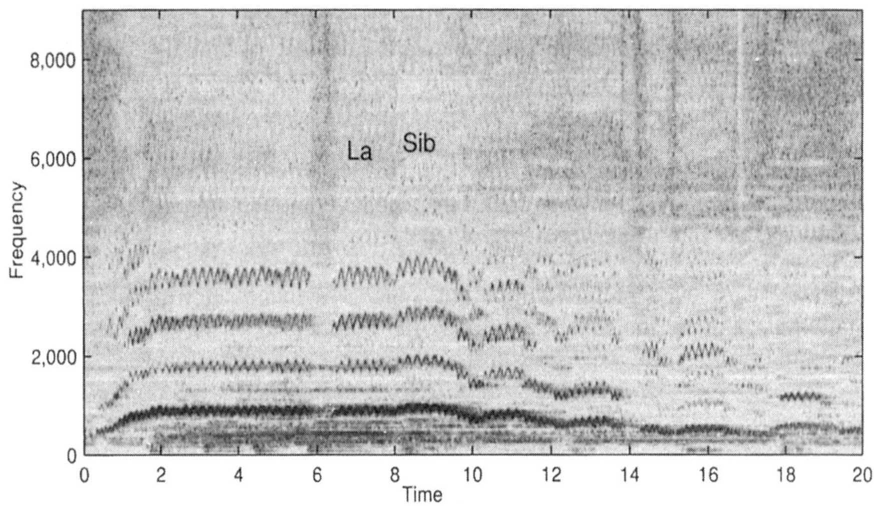

Figure 1.18. *Farther down, the famous rise*
'B C D E F G A, A B♭, A, G, F, G...'

that sounds like a continuous hiss, or static on a radio not tuned to any station. The 'color' of a noise depends on the way the values of $s(t)$ were randomly chosen.

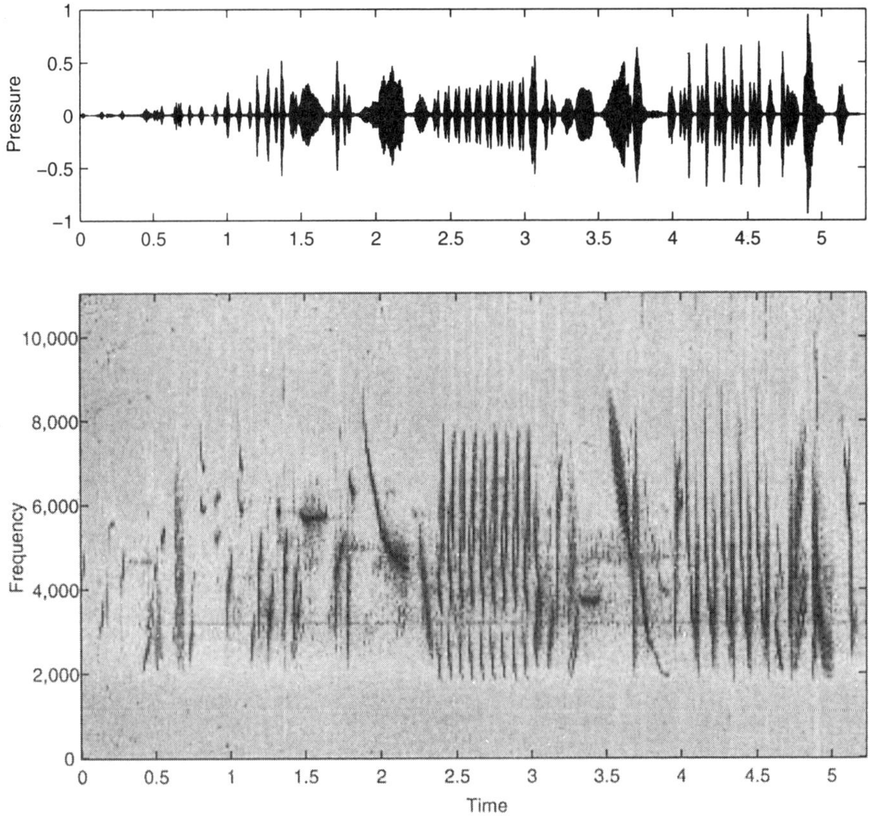

Figure 1.19. *Our thistle-eating friend treating us to a small bucolic touch. A scene that could be described as... pastoral! Pressure according to time (top); intensity in the time-frequency representation (bottom)*

White noise (by analogy with white light) is a sound that contains all audible frequencies with the same intensity. For example, the sound

$$s(t) = \sum_{n=1}^{N} \cos(2\pi f_n t + \theta_n)$$

will produce perfectly suitable white noise for a high enough value of N (several hundreds), if the frequencies f_n are chosen randomly with a *uniform* probability[8] from the

8. All of the interval's frequencies have the same probability (or the same probability density).

interval [20 Hz, 20 kHz], and if the phases are picked in the same way from the interval $[0, 2\pi]$. In the case of *pink noise*, $\log f_n$ is chosen instead randomly and with uniform probability from the interval [log 20, log 20 000], so as to have the same power (the intensity) inside every octave.

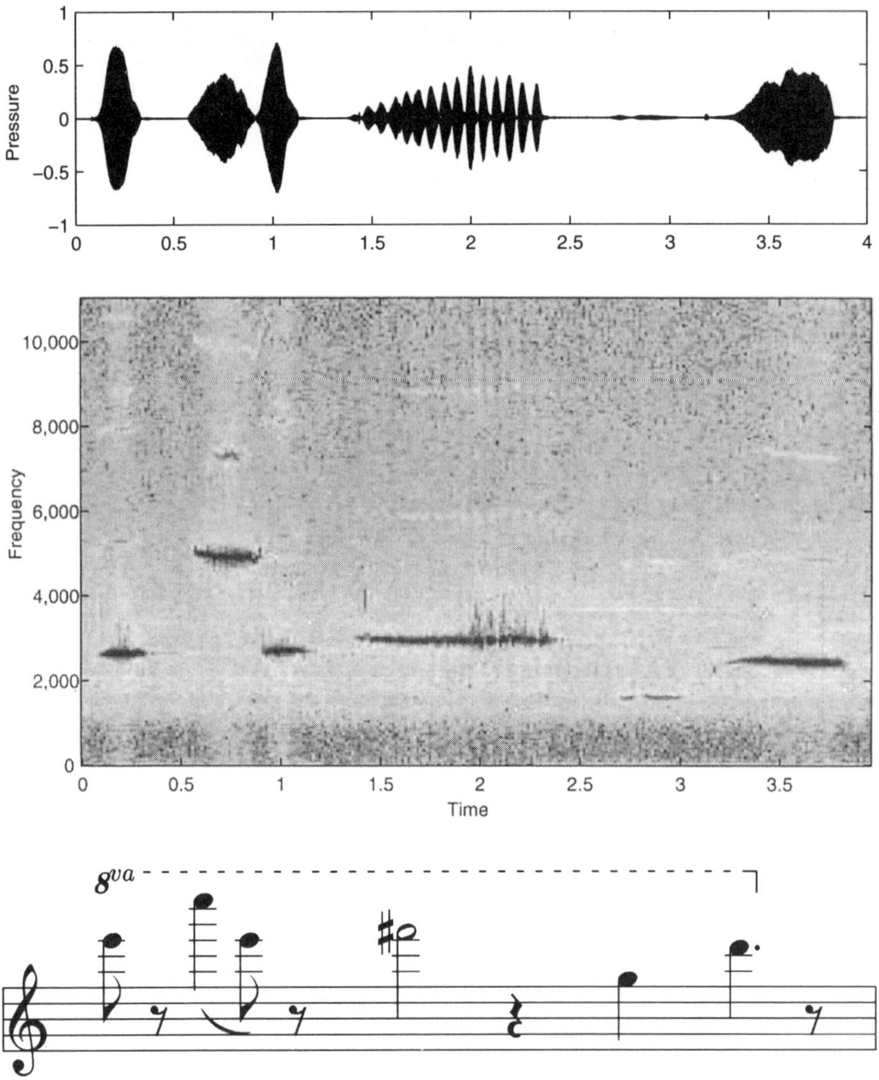

Figure 1.20. *The mountain whistler: less picturesque, maybe, but much more melodious. Pressure according to time (top); intensity in the time-frequency representation (middle); transcription* à la *O. Messiaen (bottom): do you agree with the pitches of the notes? (See the* AM *website for audio, and Chapter 4 about the sensation of pitch.)*

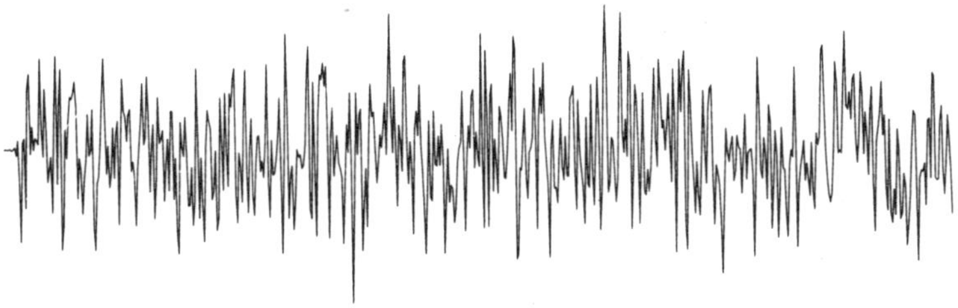

Figure 1.21. *What was that noise?*

White noise is used for example by sound engineers before a concert to detect the concert hall's resonant frequencies: these frequencies will be emphasized in the room's response to white noise. This means that, during a concert, these frequencies will be amplified compared to others. If loudspeakers are used during the concert, the sound engineer can then compensate this unwanted effect with the help of the equalizer on his mixing console, simply by attenuating the resonant frequencies with the appropriate *filters* (see section 1.5).

Finally, all sorts of noises can be obtained by filtering white noise with a *band-pass* filter (a type of filter that only passes frequencies within a certain interval). These noises can then be used for sound effects such as for example imitating the *blowing* sound produced by a wind instrument.

1.4. Representation of sound

When studying different types of sounds over the last section, we were faced with the problem of sound representation: with respect to time or frequency. Both have their pros and cons, but if we go by musical intuition, we sense the need for a representation that involves both time and frequency: this is what is called the *time-frequency representation*. Theory in this field is rich and complex. We will restrict ourselves to a simplified version, introduced with the help of a musical example. We will also see how it can be useful to speech analysis.

Suppose that the sequence of notes A, C, E, with respective frequencies 440, 523.25 and 660 Hz are played on a flute for a duration of two beats each. A musician would write Figure 1.22.

Figure 1.22. *Musical representation of the A-C-E sequence*

We will assume that each note contains only the first four harmonics, with amplitudes 64, 16, 4 and 1, respectively, and lasts one second. Let us calculate the Fourier coefficients of each note (there are four non-zero coefficients for each note), and draw the graph of the corresponding frequencies according to time, using a line with varying intensity or color depending on the amplitude. The result is a *sonogram*, or *spectrogram*, shown in Figure 1.23, first introduced in 1946 by R. Potter of Bell laboratories.

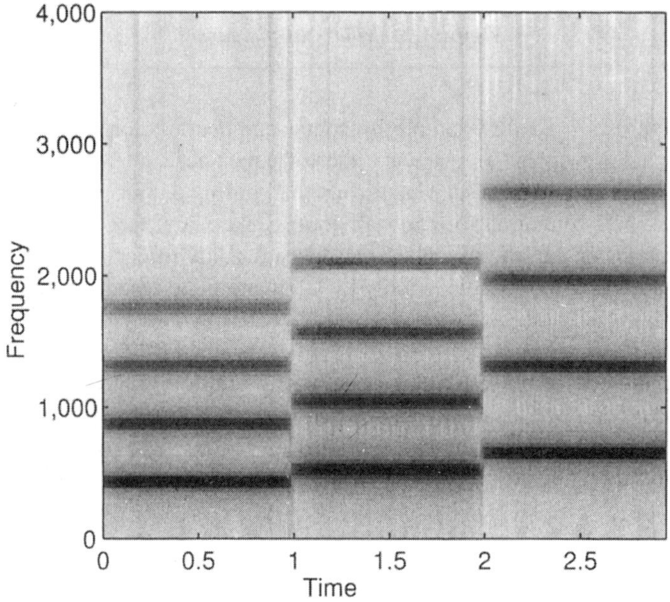

Figure 1.23. *Spectrogram of A-C-E in the case of four harmonics per note*

In this graph, the y-axis shows frequency and the x-axis shows time. The analogy with the score in Figure 1.22 is obvious, but the spectrogram carries the *additional* information of the analyzed sound's harmonics composition. A spectrogram like this one is usually not calculated using intervals of one second, as these would be much too long for sounds varying more rapidly than the one in this example. We will describe later the exact procedure used.

1.4.1. *Time or frequency analysis, discrete Fourier transform*

In phonetics, sounds fall into two categories: those that involve the *vocal chords* – vowels and sonorant consonants such as 'l' and 'm' – and those where vocal chords do not vibrate – obstruent consonants such as 'sh' and 's'. Vocal chords produce a more or less periodic sound, unlike sounds that are hissed or whistled. The three Figures 1.24, 1.25 and 1.27 show three types of speech analysis for the recorded words 'le chapeau' (the hat, pronounced luh-shah-poe). Unlike the first two, the third shows a clear distinction between the vowels and the obstruent consonants.

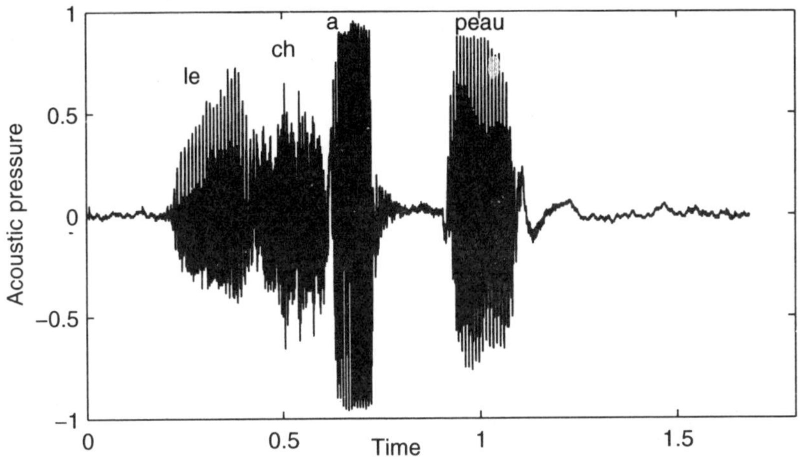

Figure 1.24. *Time representation of 'le chapeau', mixed with a faint background noise*

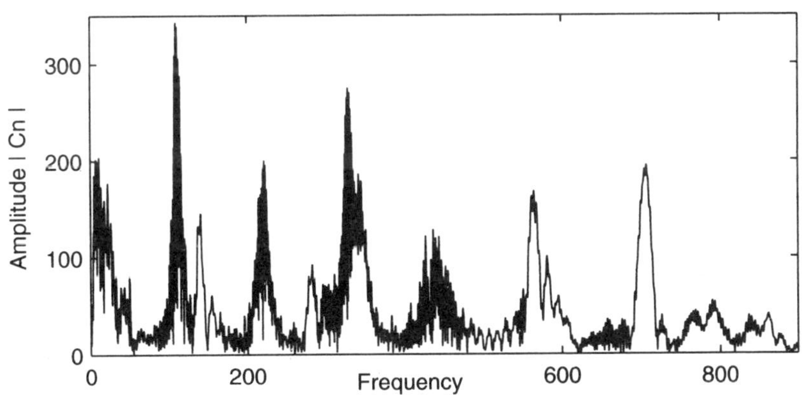

Figure 1.25. *Frequency representation of 'le chapeau'. Vowels are responsible for the peaks found at multiples of* 110 *Hz*

The first figure simply shows the time signal. The second shows the frequency analysis of the same signal. In order to do this, the recorded sound $s(t)$ is considered to be *one* period, in this case with a duration of $T = 1.7$ s, of a periodic signal (an imaginary signal outside the interval $[0, T]$, see also section 1.3.3), and then its Fourier coefficients c_n are calculated:

$$s(t) = \sum_{n=-\infty}^{+\infty} c_n \exp(2i\pi nt/T) \text{ for } 0 \leq t \leq T.$$

But the coefficients c_n were not calculated exactly from formula (1.13): first because it would not be possible to achieve such a calculation, and second because not all the values of s are known, only N values $s(t_k)$ (called samples, see Chapter 5) taken at $t_k = kT/N$. We actually performed an *approximate integration* (using rectangles to approximate the integrand) that provides the approximation

$$c_n \simeq \widetilde{c}_n = \frac{1}{N} \sum_{k=0}^{N-1} s(t_k) \exp(-2i\pi nk/N). \tag{1.16}$$

This is what is called the length N *discrete Fourier transform*[9] (DFT). Just as with the Fourier series, an inverse transformation leads from the coefficients $\widetilde{c}_n$ back to the values $s(t_k)$:

$$s(t_k) = \sum_{n=0}^{N-1} \widetilde{c}_n \exp(2i\pi nk/N). \tag{1.17}$$

The lowest frequency[10] in this case is $f_1 = 1/1.7$ Hz. The $\widetilde{c}_n$ are shown in Figure 1.25 not according to n (an integer without any physical meaning) but according to the associated frequency $f_n = n/T = n/1.7$ Hz. As you can see, certain frequencies are more intense than others, particularly near the frequencies 110 Hz, 220 Hz, 330 Hz, 440 Hz, 570 Hz, 700 Hz which approximately correspond to the harmonics of the fundamental note 110 Hz (A2) produced by the vocal chords. The drawback of this representation (Figure 1.25) is that it does not give any information regarding the sound signal's time evolution, whereas the drawback in Figure 1.24 was the lack of information on the frequencies contained in the sound. Hence the need for another kind of analysis: the *time-frequency analysis* provided by the spectrogram.

1.4.2. *Time-frequency analysis, the spectrogram*

The idea when performing the time-frequency analysis of a sound signal is as follows. We start by dividing the signal in M small segments $s_m(t)$ such that

$$s_m(t) = \begin{cases} s(t) \text{ if } t \in I_m = [mT/M, (m+1)T/M[, \\ 0 \text{ otherwise.} \end{cases}$$

---

9. A process exists to rapidly calculate all of the coefficients $\widetilde{c}_n$ in $1.5N \log_2(N)$ operations. It was invented by Cooley and Tukey in 1965 and was called the FFT (fast Fourier transform).

10. This frequency has no physical meaning, and only expresses the fact that the sound lasts 1.7 s.

Another way to express this is to set $s_m(t) = w(Mt/T - m)\, s(t)$ with

$$w(t) = \begin{cases} 1 \text{ if } t \in [0,\, 1[, \\ 0 \text{ otherwise,} \end{cases}$$

meaning that each segment $s_m(t)$ is obtained by multiplying $s(t)$ by the rectangular *sliding window* $w(Mt/T - m)$. Figure 1.26 shows a signal $s(t)$ and the third segment $s_2(t)$ for $T = 3$ and $M = 10$. Notice by the way how the frequency increases with time in signal s, like at the beginning of a siren.

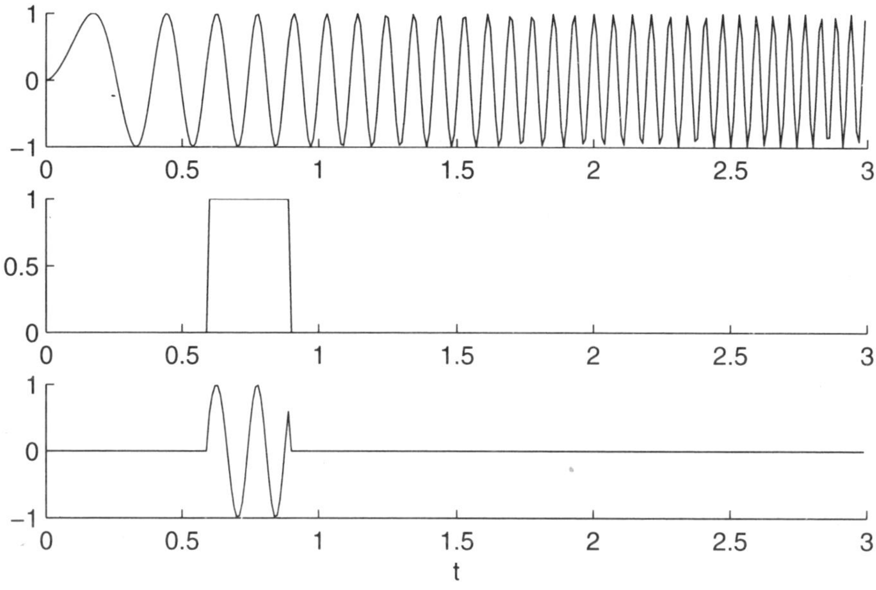

Figure 1.26. *Sound signal to be analyzed (top); rectangular sliding window (middle); multiplication of the signal by this window (bottom)*

Then, using the same idea as with the A-C-E sequence, a DFT of the signal $s_m(t)$ is performed on each interval I_m, $m = 0, 1, ..., M - 1$, and for each coefficient $\widetilde{c}_n^m$ (the index m indicating that we are dealing with the interval I_m) associated with the frequency $f_n = nM/T$, a line is drawn on the spectrogram connecting the points with coordinates $(mT/M,\, f_n)$ and $((m+1)T/M),\, f_n)$, with an intensity or color that varies with $|\widetilde{c}_n^m|$.

Typically, the length of each DFT is 256 or 512. In practice, instead of using disjoint intervals (since these can render 'invisible' what happens at the junctions, a gap for example), the intervals are chosen so that they overlap, and instead of a rectangular window, other windows are used with a less abrupt break, such as the ones shown in Figure 5.10.

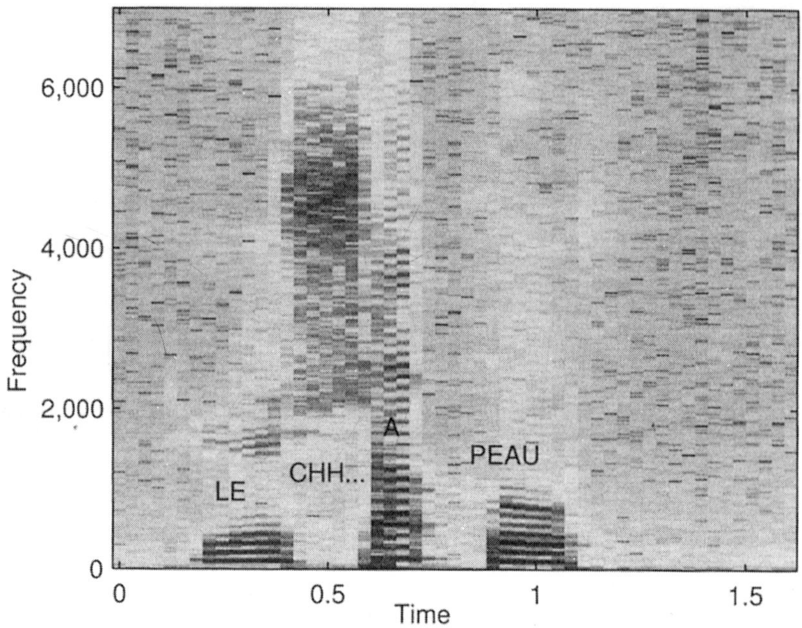

Figure 1.27. *Time-frequency representation of 'le chapeau'. The vowels are fairly 'musical', the consonant 'ch' is rather 'noisy'!*

If we apply this process to our example, we get Figure 1.27 (with length 512 DFTs and 22,050 samples per second). Notice in particular how a consonant such as 'ch' contains many more high frequencies than a vowel. We also find for the sounds 'e', 'a' and 'o' the harmonics produced by the vocal chords, that had been observed in Figure 1.25. All the same, the sound signal's evolution with time is still shown, we have information about both time and frequency.

However, notice that the image lacks sharpness. This is not due to any technical problem related to the image processing, but to a genuine impossibility of having a signal concentrated both in time and frequency, for two reasons.

The first is qualitative. For any given function $s(t)$, it is impossible for both $s(t)$ and $\widehat{s}(f)$ to be both equal to 0 outside a finite interval (except if $s = 0$). In particular, if a sound has a finite duration, then it necessarily contains arbitrarily high frequencies: for any chosen threshold f_S, this sound contains frequency components with frequencies higher than f_S!

The second, quantitative reason is known in quantum mechanics as *Heisenberg's uncertainty principle*, which states that it is impossible to measure with an arbitrary precision both the position and the speed of a particle. This principle is actually a mathematical result, according to which, for a function $s(t)$ such that

$$\int |s(t)|^2 \, dt = 1,$$

the *standard deviation*[11] σ associated with the probability density $|s(t)|^2$, and the standard deviation $\widehat{\sigma}$ associated with the probability density $|\widehat{s}(t)|^2$ are such that

$$\sigma \, \widehat{\sigma} \geq \frac{1}{4\pi}. \tag{1.18}$$

What does this actually mean for the sounds of these two properties? Imagine that the graph in our spectrogram only showed a small black or colored rectangle depending on what graphics were chosen, with the size $\varepsilon_1 \times \varepsilon_2$ corresponding to such a signal $s(t)$. This would mean that the sound's duration in time is ε_1, and that the frequencies are concentrated in an interval with length ε_2, which already contradicts the first property. But even if that were possible, we would necessarily have $\sigma \leq \varepsilon_1/2$ as well as $\widehat{\sigma} \leq \varepsilon_2/2$. Using (1.18), we would get $\varepsilon_1 \varepsilon_2 \geq 1/\pi$, the lower bound of the time-frequency 'surface' imposed on any sound!

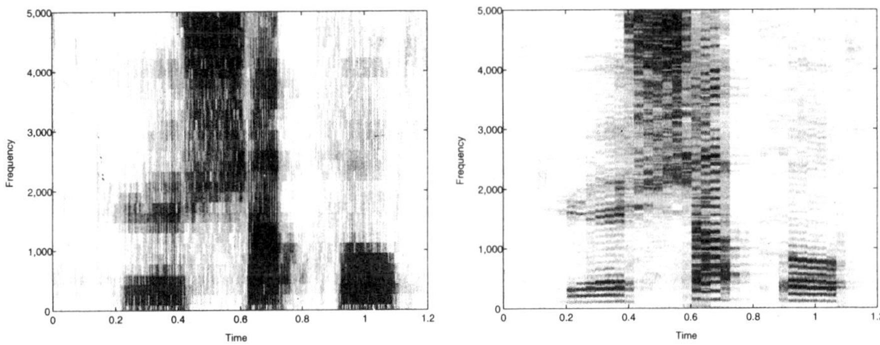

Figure 1.28. *'Le chapeau' on the left, high time resolution (128-sample window, with in this case 22,050 samples per second); on the right, high frequency resolution (1,024-sample window)*

As a result, the uncertainty principle has repercussions on the time-frequency analysis itself: the choice has to be made between high resolution for time, which leads to a loss of accuracy on frequencies, and a high resolution for frequency, which leads to a loss of accuracy on the sound's time evolution.

11. The mean associated with a density probability g is $m = \int x g(x) \, dx$, and the standard deviation σ is equal to $(\int (x-m)^2 g(x) \, dx)^{1/2}$.

This is illustrated by Figure 1.28 which shows two spectrograms of our 'chapeau', performed, in the first case, with a short time window (256 samples, or 12 ms), and in the second case with a longer time window (1,024 samples). The first time-frequency analysis is slightly better at showing the sequence of time events, whereas the second one shows more clearly the harmonics or partials produced by the vocal chords (the horizontal lines) and that were not visible in the first figure.

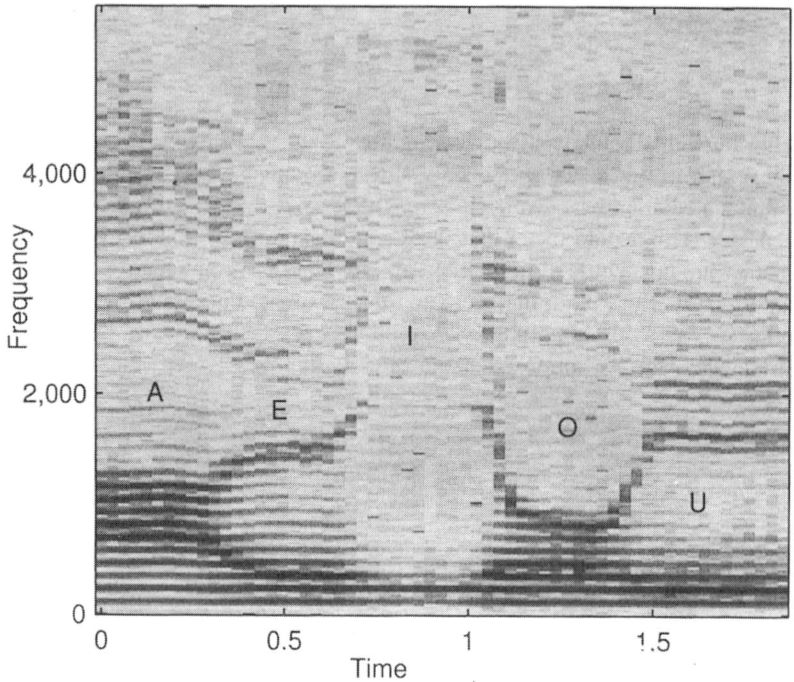

Figure 1.29. *Time-frequency representation of the vowels. Naturally, the sound 'o' leads to a circle, as any self-respecting 'o' should!*

Let us finish with a last example of a spectrogram: the analysis of 5 vowels (Figure 1.29). These sounds are fairly concentrated in the low frequencies. Also, each vowel produces darker areas, hence more intense, corresponding to the resonant frequencies of the vocal tract. Because these frequencies depend on the *form* (or shape) of the latter, they are called *formants*.

If you look closely, these formants can also be seen in Figure 1.17, where you can notice that during the first six seconds, the upper harmonics do not follow the melodic line at all (A G A C B♭ A G G) of the first two, but instead seem to just 'hover in place'. This is due to the fact that the high-pitched harmonics having the same number (at the beginning 1, 2, 6, 7, 8, 9) are not the ones that are intense in each of these notes, instead it is *those found in the* 2,500 *to* 4,000 *Hz range*. The sound is *shaped* by the vocal tract, which reinforces certain frequencies, and attenuates others, hence the name formants.

1.5. Filtering

Filtering a sound is equivalent for example to the process you perform when you turn the bass and treble dials on your stereo, resulting in a modification of the sound's low-pitched to high-pitched ratio. The following is a brief mathematical description.

1.5.1. *Discrete spectrum*

To begin, let us consider again a pure sound, consisting of only one harmonic with frequency f, written in complex form as:

$$s(t) = \alpha \exp(2i\pi ft).$$

Two basic operations can be performed on this sound. These are:

– *amplification* by a factor $a > 0$:

$$v_1(t) = as(t);$$

– *phase shifting* by an angle $\theta \in [0, 2\pi[$:

$$v_2(t) = \alpha \exp(2i\pi ft - i\theta).$$

This phase shifting can also be interpreted as a *delay* of $\tau = \theta/2\pi f$, since $2\pi ft - \theta = 2\pi f(t - \theta/2\pi f)$, and therefore

$$v_2(t) = s(t - \tau).$$

If both operations are performed, we get the sound

$$v(t) = a\alpha \exp(2i\pi f(t - \tau)) = a \exp(-2i\pi f\tau)s(t),$$

and the initial sound has been multiplied by the complex number

$$H = a \exp(-2i\pi f\tau),$$

$$v(t) = Hs(t).$$

Combining these two operations is at the core of *filtering*: amplification and/or phase shifting. Filtering is of course *linear*, and *unaffected by time shifting*, or in other words, filtering then shifting by a time interval t_0 yields the same result as a shift by the same time interval followed by the filtering:

$$\begin{aligned} s(t) &\longmapsto \quad u_1(t) = Hs(t) \quad &\mapsto u_2(t) = u_1(t - t_0) \\ s(t) &\longmapsto \quad v_1(t) = s(t - t_0) \quad &\mapsto v_2(t) = Hv_1(t). \end{aligned} \tag{1.19}$$

As expected, we have $u_2(t) = v_2(t) = Hs(t - t_0)$.

1.5.1.1. *Transfer function*

Consider now a more complex sound,

$$s(t) = \sum_n c_n \exp(2i\pi f_n t),$$

and let us apply to each term with frequency f_n the previous operation of a multiplication by a complex number dependent on the frequency, denoted by $H(f_n)$. The resulting modified sound is

$$v(t) = \sum_n H(f_n) c_n \exp(2i\pi f_n t). \tag{1.20}$$

Again, this operation is linear and unaffected by time shifting, and the system that changes s into v is called a *filter*[12]. The function $H(f)$ (considered here only for the frequencies f_n, but it can just as well be defined for any frequency f) is called the filter's *transfer function*.

What is the point of filtering? Essentially to analyze a sound signal or to modify its frequency component. If for example a sound is considered to be too 'bright', it should be filtered with a filter whose $|H(f_n)|$ are low (or even equal to zero) for the high frequencies, and close to 1 for the low frequencies. Such a filter is called a *low-pass* filter. You can also do the opposite: attenuate the low-frequencies; this is done with a *high-pass* filter. Finally, it is also possible to select the intermediate frequencies and to attenuate all the others, using what is called a *band-pass* filter. We will discuss this further later on.

For example, the roughly triangular signal from Figure 1.14 (bottom plot), comprised of the first four harmonics and with fundamental 1 Hz, was filtered using three

12. It can actually be demonstrated that any operation that is linear and unaffected by time shifting can be written in that form.

filters, the transfer functions of which are shown in the left column of Figure 1.30: a low-pass, a band-pass and a high-pass. The column on the right shows the output signals. For the first two, these filterings resulted in singling out the components with frequency 1 Hz and 2 Hz respectively. The third filter isolated the sum of the two components with frequencies 3 Hz and 4 Hz.

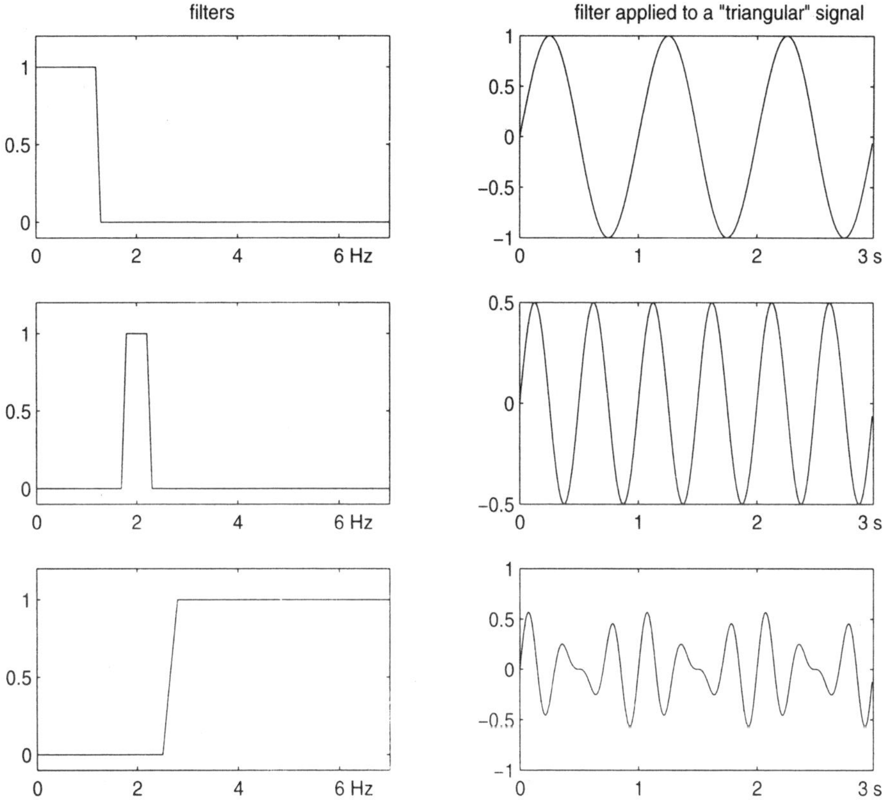

Figure 1.30. *Different filters applied to a 'triangular' signal*

Figure 1.31 shows the same process repeated on the roughly rectangular signal from Figure 1.15, again with the fundamental 1 Hz. Can you interpret it?

1.5.1.2. *Impulse response*

Furthermore, it can be shown that if certain conditions are met (for example $H(f)$ and $\widehat{H}(f)$ are integrable), $H(f)$ is the Fourier transform of a function $h(t)$:

$$H(f) = \widehat{h}(f). \tag{1.21}$$

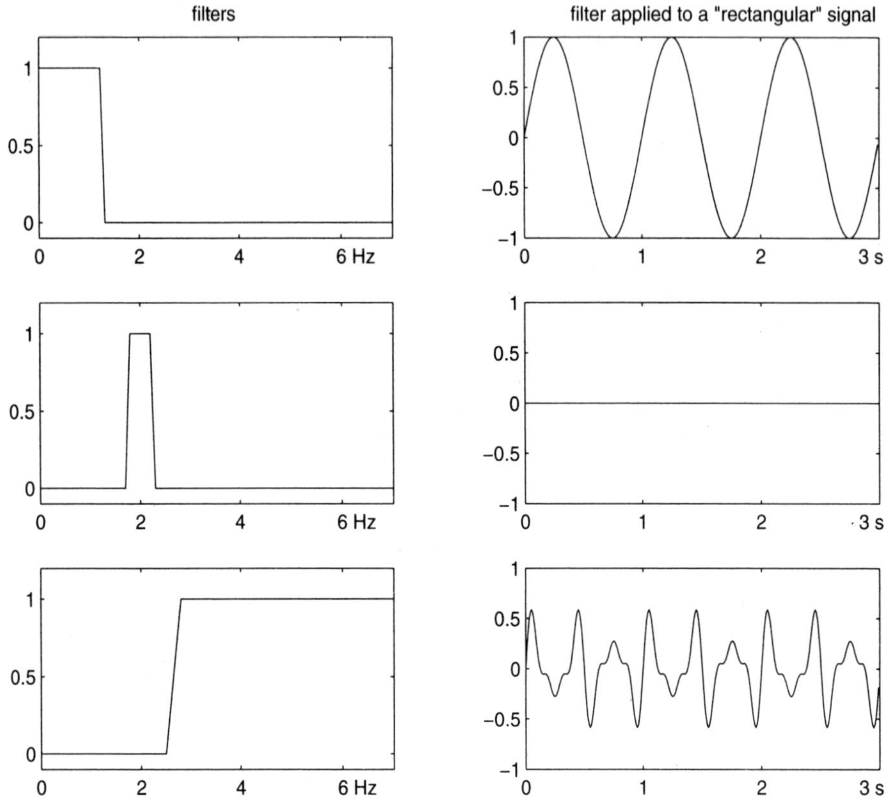

Figure 1.31. *Different filters applied to a 'rectangular' signal. What happened?*

This function $h(t)$ is called the filter's *impulse response*: this is the filter's output signal when the input is the *Dirac impulse* in 0, discussed in greater detail in section 5.1. This impulse, denoted by $\delta(t)$, is an infinitely brief signal (but it is not an actual function!), with all its energy focused in 0, and such that

$$\int_{-\infty}^{+\infty} \varphi(t)\delta(t)\,dt = \varphi(0) \qquad (1.22)$$

for any *continuous* function φ, that is to say, a function without any 'gaps'.

The function v resulting from the filtering given by formula (1.20) can then be expressed as the *convolution product* of h and s:

$$v(t) = (h * s)(t) = \int_{-\infty}^{+\infty} h(t-u)s(u)\,du. \qquad (1.23)$$

This product obeys the following important property, called commutativity:

$$h * s = s * h.$$

1.5.2. *Continuous spectrum*

Let us now go over the filtering of sounds with continuous spectra, which, if you remember, means sounds of the form

$$s(t) = \int_{-\infty}^{+\infty} \widehat{s}(f) \exp(2i\pi ft)\, df$$

where $\widehat{s}$ is the Fourier transform of s (see equation (1.15)).

Just as with the examples studied previously, filtering a sound signal $s(t)$ using a filter with the transfer function $H(f)$ will yield the signal $v(t)$, the Fourier transform of which will be $H(f)\widehat{s}(f)$:

$$\widehat{v}(f) = H(f)\widehat{s}(f). \tag{1.24}$$

Thus the value of the transfer function $H(f)$ can be seen as the amplification factor/phase shift of the signal $s(t)$ at the frequency f.

The function $H(f)$ is the Fourier transform of a function $h(t)$, still called the impulse response, and again we have the convolution product

$$v = h * s. \tag{1.25}$$

The underlying mathematical theory was developed in the 19th and early 20th centuries, particularly by Laurent Schwartz (1915-2002), who 'invented' distributions, a generalization of the concept of functions (see for example [GAS 90]).

Application: the sound received by a listener in a concert hall can be seen as the result of the sound coming from the orchestra, filtered by the filter made up of the room itself. A rough idea of its impulse response can be gathered from clapping one's hands or emitting a very brief sound. What do you think the listener will perceive if the room's transfer function resembles the graph shown in Figure 1.32?

1.5.3. *Ideal low-pass, band-pass and all-pass filters*

This chapter ends with the description of the models for three fundamental filters, which we will be using several times.

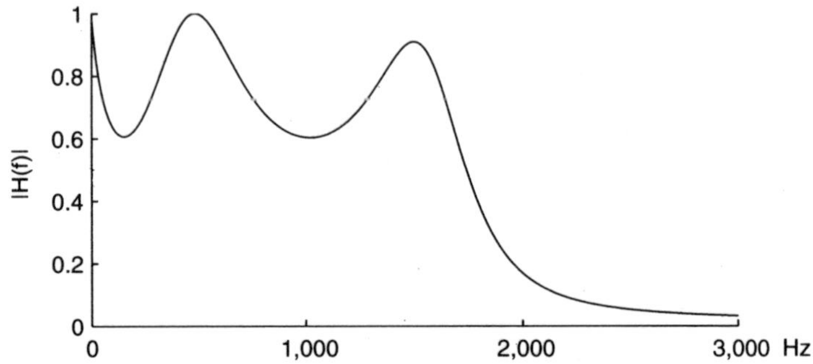

Figure 1.32. *Transfer function of an imaginary concert hall*

The *ideal low-pass filter*[13] with *cut-off frequency* $B > 0$ (see Figure 1.33) is given by its transfer function[14]:

$$H(f) = \begin{cases} 1 \text{ if } |f| < B \\ 0 \text{ otherwise.} \end{cases} \tag{1.26}$$

Hence this filter passes the frequencies $|f| < B$ without affecting them and blocks the frequencies $|f| > B$ (nothing can be said of the limit case $|f| = B$).

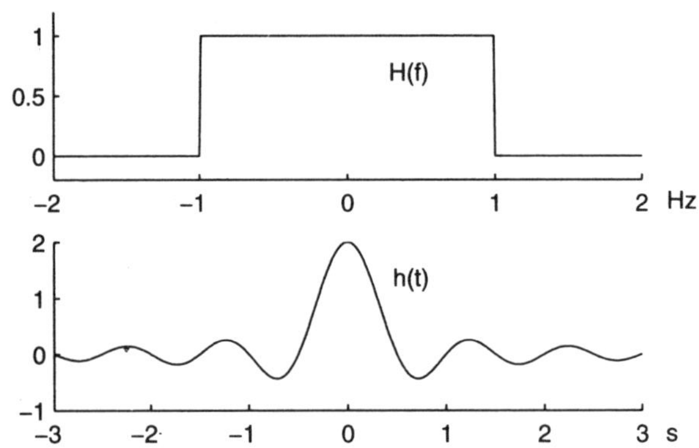

Figure 1.33. *Transfer function (top) and impulse response (bottom) of the ideal low-pass filter with cut-off frequency* $B = 1$

13. It is called 'ideal' because it cannot be achieved physically. It can merely be approximated as precisely as desired by an electronics system, but only by allowing a certain delay for the output.

14. The value of H at both ends B and $-B$ has no importance in theory, since the integral does not 'see' isolated punctual values, except if there are Dirac masses at those points, but that is another story...

If we remember that $H(f) = \widehat{h}(f)$, the impulse response h can be obtained by using (1.14):

$$h(t) = \int_{-B}^{B} 1 \exp(2i\pi ft)\, df = \frac{\sin(2\pi Bt)}{\pi t}.$$

Hence, we have

$$h(t) = 2B \operatorname{sinc}(2Bt),$$

where the *sine cardinal* function ('cardinal' because it is equal to zero for all *integers* $\neq 0$) is given by

$$\operatorname{sinc}(t) = \frac{\sin(\pi t)}{\pi t}.$$

The *ideal band-pass filter* with *cut-off frequencies* $f_0 - B > 0$ and $f_0 + B$ (see Figure 1.34) is given by its transfer function:

$$H(f) = \begin{cases} 1 \text{ if } |f \pm f_0| < B \\ 0 \text{ otherwise.} \end{cases} \tag{1.27}$$

Hence this filter leaves unaffected the intermediate frequencies f such that $|f \pm f_0| < B$, and blocks all other frequencies.

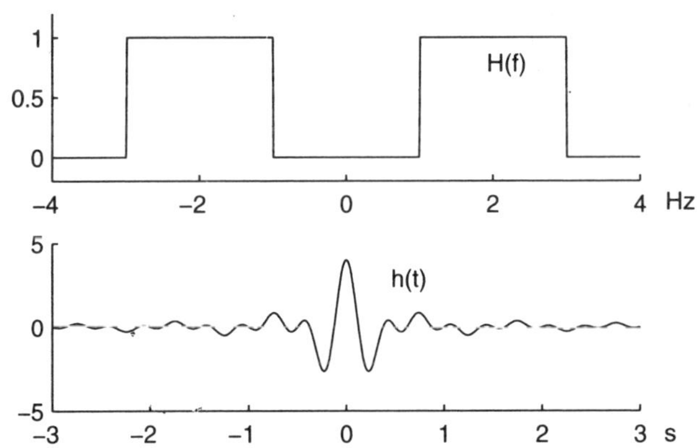

Figure 1.34. *Transfer function (top) and impulse response (bottom) of the ideal band-pass filter with cut-off frequencies 1 and 3*

The calculation of its impulse response yields

$$h(t) = 4B \operatorname{sinc}(2Bt) \cos(2\pi f_0 t).$$

Notice that this is a frequency $\cos(2\pi f_0 t)$ (called a *carrier wave* in radio communications) with its amplitude modulated by the impulse response of the ideal low-pass filter.

Finally, the *all-pass filter*, with its strange name, passes everything! The modulus of its transfer function is 1, hence it is of the form

$$H(f) = \exp(-i\theta(f)), \quad \theta(f) \in \mathbb{R}.$$

A pure sound $\exp(2i\pi ft)$ passing through this filter is transformed into $\exp(2i\pi ft - i\theta(f))$: it therefore sees its phase shift by an angle $\theta(f)$, dependent on the frequency, but no amplitude modification. This kind of filter is used to simulate reverberation (see Chapter 6).

1.6. Study problems

1.6.1. *Normal reflection on a wall (*)*

In the half-space $x \geq 0$, with coordinates x, y, z, an incident plane wave $p_i(x,t) = \sin(kx + 2\pi ft)$ is reflected by a wall. This wall, with equation $x = 0$, is assumed to be perfectly rigid. Therefore the speed of air is zero in $x = 0$. The incident plane wave and the reflected plane wave $p_r(x,t) = \beta \sin(kx - 2\pi ft)$ produce an acoustic pressure

$$p_a(x,t) = p_i(x,t) + p_r(x,t).$$

1) Using the Euler equation, show that $\partial_x p_a(0,t) = 0$ for any t.

2) Calculate the value of β and show that $p_a(x,t) = 2\cos(kx)\sin(2\pi ft)$. What is the nature of this plane wave?

1.6.2. *Comb filtering produced by a microphone located near a wall (**)*

While recording a pure sound with frequency f, generating a harmonic acoustic wave, a microphone is placed close enough to a wall that the produced wave can be considered a plane wave in that spot. Hence the conditions of Study problem 1.6.1 are satisfied.

1) What will the sound intensity be at a given point over a period $T = 1/f$?

2) Let us assume that the microphone, placed at a distance d from the wall, reacts only to variations in pressure. For what frequency values would the signal's amplitude $\alpha(f)$ detected by the microphone reach its maximum? Its minimum? What effect does the distance d have on these values? Make a plot of $10\log(\alpha^2(f))$.

3) Same question assuming that the microphone only reacts to variations in speed (read [JOU 00] to learn more about microphones).

1.6.3. *Summing intensities (***)*

A listener is located far enough from the orchestra that the 10 violins can be considered to produce, each on their own level, a traveling plane pressure wave

$$p_i(x,t) = u_i(x - ct), \quad i = 1, 2, \ldots, 10.$$

1) Use the Euler equation and the state equation to show that the corresponding air speed is $v_i(x,t) = p_i(x,t)/c\rho_0$.

The listener is located at a fixed point x. Each violin plays the same note, with fundamental f, with the same strength, so that

$$p_i(x,t) = s(t - \varphi_i),$$

where the φ_i reflect the possible phase differences between the sounds at point x, and where the function s is T-periodic with $T = 1/f$.

2) First, calculate according to s the sound intensity I_1 produced at this point by a single violin over a period T.

3) By what factor would we have to multiply the amplitude of this violin in order to get an intensity increase of 10 dB? 20 dB? (Answer: by $\sqrt{10} \simeq 3.16$; by 10.)

4) All 10 violins are now playing together. At point x, the total acoustic pressure and the associated speed are therefore

$$p_a(x,t) = \sum_{i=1}^{10} p_i(x,t), \quad v(x,t) = \sum_{i=1}^{10} v_i(x,t).$$

Show that thc total intensity I_{10} at point x is

$$I_{10} = \frac{1}{c\rho_0 T} \int_0^T \left(\sum_{i=1}^{10} s(t - \varphi_i) \right)^2 dt.$$

With the help of the Cauchy-Schwarz inequality

$$\int_0^T g(t)h(t)\,dt \leq \left(\int_0^T g^2(t)\,dt \right)^{1/2} \left(\int_0^T h^2(t)\,dt \right)^{1/2},$$

use this result to show that $0 \leq I_{10} \leq 100 I_1$, and that the associated decibel levels are such that $L_{I_{10}} \leq L_{I_1} + 20$ dB.

5) In what situation do we have $L_{I_{10}} = L_{I_1} + 10$ dB as mentioned in this chapter?

1.6.4. *Intensity of a standing wave (**)*

Consider a standing pressure wave, of the form

$$p_a(x,t) = a\sin(k(x-x_0))\cos(2\pi f(t-t_0)).$$

Using the Euler equation to determine the speed v, show that the intensity over a period $T = 1/f$ is equal to zero. Interpretation: a standing wave carries no energy (it only fluctuates without traveling).

1.6.5. *Sound of a siren (*)*

The siren was invented by the French engineer Cagniard de La Tour (1777-1859). For a sound of the form $s(t) = \sin(2\pi F(t))$, the function $f(t) = F'(t)$ is called the *instantaneous frequency*. Determine the expression $s(t)$ of a siren whose instantaneous frequency is a sinusoidal variation, except for a constant, between the two frequencies $f_0 - \beta$ and $f_0 + \beta$.

1.7. Practical computer applications

This first series of practical applications deals with synthesizing, listening to, and analyzing sounds using the MATLAB software. The necessary programs and sound files can be found on the `AM` website, in the `TP` folder. You will also find certain answers in the `TP/CORRIGES` folder. The address of the `AM` website is:

`www-gmm.insa-toulouse.fr/~guillaum/AM/`

It is recommended that you copy the entire content of the `TP` folder to your working folder, meaning the folder you will be using as the default directory during a MATLAB session. This will allow you to edit and modify them as you please.

Typographic convention: mathematical objects are written in *italic* (for example the sound $s(t)$). MATLAB objects are written in `typewriter` style (for example the third element `s(3)`).

1.7.1. *First sound, vectors*

On a computer, a sound is represented by its values $s_n = s(t_n)$, called *samples*, with $t_n = n\tau$ and $\tau = 1/F_e$. Thus two consecutive instants t_n and t_{n+1} are separated by a time interval τ called the *sampling period*, and the number of samples per second, equal to F_e, is called the *sampling frequency* (these concepts are discussed in detail in Chapter 5).

In MATLAB, the values s_n can be arranged in a vector `s`, and the element number n can be accessed by typing `s(n)`. An example of the creation of a vector and of the access to one of its elements: after starting MATLAB, write the following lines (the ≫ sign is the 'prompt' that shows up in MATLAB), hitting the 'enter' key at the end of each line:

```
≫ s = [1,-0.5,2,3];
≫ s
≫ s(3)
```

Notice that the result of the operation *is not displayed* or *is displayed* depending on whether the line ends with a semi-colon (line 1) or not (lines 2 and 3).

A vector $x = [a,\ a{+}h,\ a{+}2h,\ ...,\ a{+}nh]$ with its elements at equal distances from each other can easily be created: simply execute the command `x = a:h:(a+n*h);` after assigning values to the variables `a`, `h` and `n`.

First sound (pure sound): analyze and execute the following command lines:

```
≫ Fe = 22050;
≫ f = 440;
≫ T = 1;
≫ dt = 1/Fe;
≫ t = 0:dt:T;
≫ s = sin(2*pi*f*t);
≫ sound(s,Fe);
```

An essential tool: the online help. For a brief description of a MATLAB function, simply execute the `help` command followed by the function's name, for example `help sin` or `help sound`. And if you need help with the help, type `help help`! One of the common features of most MATLAB functions is to return a vector if the argument itself is a vector. For example, the vector `s` above is comprised of the values $s(t_n) = \sin(2\pi f t_n)$ in the interval $[0, T]$ that was specified.

1.7.2. *Modifying the parameters: the command file*

If we wish to modify the frequency for example, it would be tedious to write everything over again. Instead, we should use a file containing all the commands. Open a file, name it `test1.m`, and write the previous list of commands in this file (without the ≫!). To execute *all of the commands* contained in this file, simply type the command `test1` at the MATLAB prompt, after saving the file of course.

Now modify the values of Fe, T and f to your liking in the test1.m file, and interpret what you hear when you run the file.

1.7.3. *Creating more complex sounds: using functions*

We now want to create a more complex sound that contains several frequencies. You may keep writing in the previous command file, but it is more convenient to use a *function* if we wish to perform different tests.

Unlike the command file, functions return the results of calculations that require one or several arguments. These functions are also written in a file with the .m extension, but the first line, called the *header*, must be written with the following structure

```
function [y1,y2,...,yp] = funct(x1,x2,...,xq)
```

where funct.m is the file's name.

Open and read the synthad.m file that you should have copied from the AM website to your current directory. The lines starting with % are comments (this means that they are not read by the MATLAB interpreter).

Execute for example the following commands (the other arguments keep their previous values):

```
>> a = 1; p = 0;
>> s = synthad(a,f,p,T,Fe);
>> sound(s,Fe);
```

This function allows us to generate more complex sounds: if we give synthad the vectors $a = [\alpha_1, \alpha_2, ..., \alpha_m]$, $f = [f_1, f_2, ..., f_m]$, $p = [\theta_1, \theta_2, ..., \theta_m]$ and the numbers T, F_e as its arguments, the function will return in the vector s the samples $s(t_n)$ of the sound

$$s(t) = \sum_{n=1}^{m} \alpha_n \sin(2\pi f_n t + \theta_n), \quad 0 \leq t \leq T.$$

A few technical points before we go any further:

Vector operations. Transposing a line vector into a column vector, or vice versa, is done by adding a right quote: x'. Adding or substracting vectors of the same size is done using the $+$ and $-$ operators. A bit more odd: the sum a+x of a number a and a vector x adds a to each component of x. Multiplying and dividing a vector x by a number a is done by writing a*x and x/a. MATLAB offers convenient tools to perform operations on vectors without resorting to the use of loops, the .* and ./

operators, which work *term by term*. To raise all of the terms of a vector x to the power m, type x.^m. As an illustration, type in the following commands:

```
≫ x = [1,2,3]; y = [2,2,3];
≫ x
≫ x'
≫ x+y
≫ x'+y
≫ x+0.1
≫ 3*x
≫ x/2
≫ x.*y
≫ x./y
≫ x.^2
≫ x.^y
```

The line x'+y caused an error! Two vectors can only be added if they are of the same type: line or column. Now it is your turn to change the values of the amplitudes and the frequencies given to the synthad function, and to compare the resulting sounds. In the harmonic case, you can, in particular, play around with the decrease speed of the coefficients α_n, or with the presence or absence of even harmonics. In the case of partials, try using the values given for a bell in section 1.3.

1.7.3.1. *Noise and siren interlude*

Based on synthad.m, create a function called noise.m with the header

```
function s = noise(T,Fe)
```

that creates a noise either by using the models described in this chapter or simply by using the MATLAB function randn (think of help). Create a function called siren.m as well, with the header

```
function s = siren(f,f1,beta,T,Fe)
```

that returns the samples of the sound

$$s(t) = \sin(2\pi f t + \beta \sin(2\pi f_1 t)/f_1).$$

1.7.4. *Analysis*

You are now going to analyze a sound of your choice, after creating it or copying it from the AM website (those are the files with the .wav extension). In order to read the file flute.wav for example, execute the command

```
[s,Fe] = wavread('flute.wav');
```

The variable s then contains the sampled sound and Fe contains the sampling frequency. To listen to this sound, use the sound command. Note that the vector s comes out as a column vector.

Below is an example of a sound consisting of three consecutive notes, each of which comprises four harmonics (we will no longer show the prompt ≫):

```
a = [1 000,100,10,1];
f - [440,880,1320,1760];
p = [0,0,0,0]+pi/2;
T = 1; Fe = 11025;
s1 = synthad(a,f,p,T,Fe);
s2 = synthad(a,1.5*f,p,T,Fe);
s3 = synthad(a,2*f,p,T,Fe);
s = [s1,s2,s3];
soundsc(s,Fe);
```

The second line from the bottom yields a vector `s`, resulting from writing the three vectors `s1,s2,s3` one after the other. In this case, we used the `soundsc` function rather than `sound` which only performs properly for values taken from the interval $[-1, 1]$. The `soundsc` function (`sc` for 'scale') automatically multiplies and shifts the sound samples so as to have the minimum equal to -1 and the maximum equal to 1. It has a drawback however: it does not necessarily render a sound starting or ending near 0, which can cause a 'click' at the start or at the end of the sound. This inconvenience can be avoided by writing `s = s/max(abs(s)); sound(s,Fe);`

1.7.4.1. *Time analysis*

Time analysis is the time representation of the signal, performed with the `plot(x)` command. You can zoom in by clicking on the ⊕ icon and choosing the area to enlarge with the left mouse button, or zoom out with the right mouse button.

1.7.4.2. *Frequency analysis*

To perform the frequency analysis of a sound $s(t)$ with a duration T, on the frequency range $[0, F_{\max}]$ where the condition $F_{\max} < F_e/2$ is imposed (the explanation will come in Chapter 5), run the following command lines. In this case, we chose $F_{\max} = 4,000$:

```
z = fft(s);
T = (length(s)-1)/Fe;
fr = 0:1/T:4000;
nf = length(fr);
plot(fr,abs(z(1:nf)));
```

If for example you chose for `s` a periodic sound with the harmonics 440 Hz, 880 Hz and 1,320 Hz, you should see three peaks appear for the x-coordinates 440, 880 and 1,320, with heights proportional to the weights (the α_n) of each of these harmonics.

1.7.4.3. *Time-frequency analysis*

The time-frequency analysis is performed using the MATLAB function `specgram`:

```
specgram(s,512,Fe);
```

The number 512 indicates the width of the analysis window (see section 1.4.2), and the calculation speed of the FFT is optimum if that number is a power of 2. Try different values and interpret the resulting spectrograms.

1.7.5. *Filtering*

We are going to filter the sound `steinwayE.wav` that you copied from the `AM` website. Open it in MATLAB by executing the command

```
[s,Fe] = wavread('steinwayE.wav');
```

We will admit that the following lines perform a *low-pass* filtering with cut-off frequency $W = W_n * F_e/2$ where $W_n \in]0,1[$:

```
b1 = fir1(100,Wn);
s1 = filter(b1,1,s);
```

Change the value of `Wn` so as to have W vary between 100 Hz and $1,000$ Hz and listen to the result.

Likewise, we will admit that the following lines perform a *high-pass* filtering with the same cut-off frequency $W = W_n * F_e/2$:

```
b2 = fir1(100,Wn,'high');
s2 = filter(b2,1,s);
```

Again, change the value of W in the previous interval and listen to the result.

Finally, perform a band-pass filtering of your choice by consulting the online help for the function `fir1`.

Chapter 2

Music Instruments

After studying the nature of musical sounds and their propagation, we are now going to focus on the sources of these sounds: music instruments. Composers, musicians and scientists have always been interested in understanding the production mode of musical sounds. Jean-Philippe Rameau used to say for example that 'the mere resonance of the sonorous body accounts for all of music theory and practice' [RAM 37]. Numerous mathematicians and physicists have brought their contributions [FIC 96]. However, we are far from having complete models at our disposal to describe music instruments exactly. The way they function can be extremely complex, and require the use of very sophisticated theories. Turbulence theory is needed, for example, to describe the oscillations of the flow of air produced at the mouthpiece of a flute or an organ pipe [INS 95, FLE 98].

Simply put, a music instrument is made of two essential parts: the *vibrator* (the source of the vibrations) and the *resonator*. A string by itself hardly produces any sound. It needs to be combined with a resonator to more efficiently transform the vibration's mechanical energy into acoustic energy. This may have been discovered in prehistoric times when people used their mouths when pulling the string on a bow: 10,000 to 15,000 year old cave art found in the *Trois-Frères* cave, in Ariège, France, shows a sorcerer holding the upper part of a bow between his teeth, the lower part in his left hand, and playing the string with his right hand. In most string instruments, the resonator is a sheet of spruce wood called the *sounding board*, connected to the string by a piece of wood (beech or maple) called the *bridge*. In wind instruments, the vibrator can be a reed (clarinet, saxophone, oboe, etc.), the musician's lips (horn, trumpet, trombone, etc.), a flow of air (flute, organ, etc.), and the resonator is the column of air surrounded by the instrument, sometimes the pipe itself, depending on the material it is made of.

The objective of this chapter is to study simple vibration models of sonorous bodies, the basic mechanisms of all music instruments. These vibrations can be divided into two categories: *free* vibrations and *driven* vibrations. Percussion instruments, the harpsichord, the piano, the guitar, but also the violin when playing *pizzicato*, belong to the first category. The vibrations are said to be free because after a brief action (percussion, plucking), the body is no longer subjected to an external constraint, and continues to vibrate on its own. Wind instruments (woodwind, brass, organ) and bowed instruments (violin, cello, double bass) belong to the second category. The sound is sustained by the musician blowing into the mouthpiece for wind instruments, or by the bow in the case of string instruments. We will study free vibrations of strings, bars and membranes, then driven oscillations in a tube. Air coupling will be studied in the particular case of an infinite board, and is the subject of Study problem 2.6.7. Finally, we will see how the different mechanisms described induce the instrument's characteristic property: its timbre.

2.1. Strings

2.1.1. *Free vibrations of a string*

Let us examine a string with length L and mass per unit length μ bound at both ends and subjected to a tension T[1]. We will ignore the interaction with the bridge and the important resulting attenuation; these will be the subject of Study problem 2.6.5. We will also ignore other sources of attenuation, for which you may refer to [VAL 93]. There are three possible types of vibrations: transverse, longitudinal and torsional vibrations. We will only be studying the first kind, and we will assume that the vibration occurs in an xOy plane. In this plane, the string's ends are located at points $(0, 0)$ and $(0, L)$, and the string's position at a time t is given by the equation $y = u(x, t)$. The *boundary conditions* (at the ends of the string, where it is bound) impose that $u(0, t) = u(L, t) = 0$ for any t.

In order to obtain the equation that governs the string's movement, we have to consider, for a given time t, the forces applied to a small portion of the string located between the x-coordinates x and $x+dx$ (see Figure 2.1). The angle between the string and the Ox axis is denoted by $\theta(x, t)$. At point $x + dx$, the vertical component of the force due to tension is

$$F(x + dx, t) = T \sin\theta(x + dx, t) \simeq T \tan\theta(x + dx, t) = T\partial_x u(x + dx, t),$$

1. If we denote the cross-section area and the elongation of the string by S and ΔL, we have $T = SE\Delta L/L$ where E is the elastic modulus, also called the Young modulus, of the material used for the string (roughly $2 10^{11}$ Pa for steel, with 1 Pa $= 1$ Nm^{-2}).

where the approximation is valid if $\theta(x,t)$ is close to 0. At point x, we have, likewise,

$$F(x,t) \simeq -T\partial_x u(x,t).$$

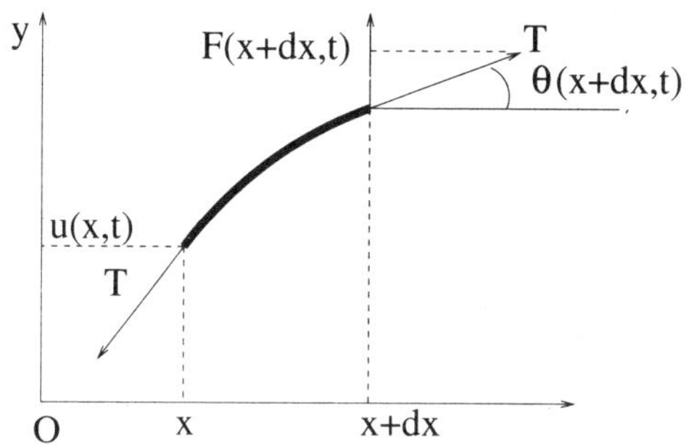

Figure 2.1. *A portion of a string subjected to tension*

Newton's second law of motion $F = m\gamma$ yields the vertical acceleration

$$T\partial_x u(x+dx,t) - T\partial_x u(x,t) \simeq \mu dx \partial_{t^2} u(x,t).$$

If we divide by dx and if dx tends to 0 (calculation of the derivative with respect to the first argument), we again obtain the wave equation (1.6)

$$c^2\partial_{x^2}u(x,t) = \partial_{t^2}u(x,t),$$

where

$$c = \sqrt{T/\mu} \tag{2.1}$$

is now (see analysis conducted in section 1.1.2) the propagation speed of a transverse traveling wave propagating along the string (not to be confused with the speed of a point on the string). As we did in Chapter 1, we can start by searching for the harmonic solutions

$$u(x,t) = \varphi(x)\exp(2i\pi ft)$$

where $\varphi(x)$ is a solution to the Helmholtz equation (with $k = 2\pi f/c$)

$$\varphi''(x) + k^2\varphi(x) = 0. \tag{2.2}$$

The solutions to this equation are of the form

$$\varphi(x) = \alpha\exp(ikx) + \beta\exp(-ikx).$$

Taking into account the boundary conditions $\varphi(0) = \varphi(L) = 0$, referred to as the Dirichlet conditions, yields the following linear system with unknowns α and β:

$$\begin{cases} \alpha + \beta = 0, \\ \alpha \exp(ikL) + \beta \exp(-ikL) = 0, \end{cases}$$

which can only have a non-zero solution if the determinant $\exp(-ikL) - \exp(ikL) = -2i \sin(kL)$ is equal to zero, that is to say, if[2]

$$k = k_n = \frac{n\pi}{L}, \quad n \in \mathbb{Z}, n \neq 0,$$

$$f = f_n = \frac{nc}{2L}. \tag{2.3}$$

If such is the case, we then have $\alpha = -\beta$ and $\varphi(x)$ is proportional to $\sin(k_n x)$. Knowing that $\exp(ikx) - \exp(-ikx) = 2i \sin(kx)$, we get a set of stationary waves

$$u(x,t) = a \sin(k_n x) \exp(2i\pi f_n t).$$

The first frequency is the fundamental f_1, and the other frequencies f_n are *integer* multiples of f_1: hence they are harmonics. Theses frequencies are called the string's *natural frequencies* and the corresponding harmonic waves are called the *natural modes*.

The first three natural modes are shown in Figure 2.2. In particular, equation (2.3) tells us that the frequency is inversely proportional to the length of the string, which had already been discovered by Pythagoras (500 BC), who, based on this observation, built the diatonic scale from a cycle of fifths (see Chapter 3).

We can then show that any solution to the wave equation for the string bound at both ends can be obtained from a sum of natural modes (see also 2.6.1 for another resolution technique).

Thus, the general form describing the string's position, a form we owe to Bernoulli (1753, [ESC 01]), is as follows:

$$u(x,t) = \sum_{n=-\infty}^{+\infty} c_n \sin(k_n x) \exp(2i\pi f_n t). \tag{2.4}$$

Because the frequencies are integer multiples of f_1, it follows that the produced sound is *periodic*, with period $1/f_1$. Reality is actually a bit different from this simplified model, as we will see at the end of this chapter when we discuss timbre.

2. The case where $n = 0$ is discarded since it leads to a zero solution $u = 0$. Note that negative frequencies turn up as positive frequencies when we switch back to real numbers.

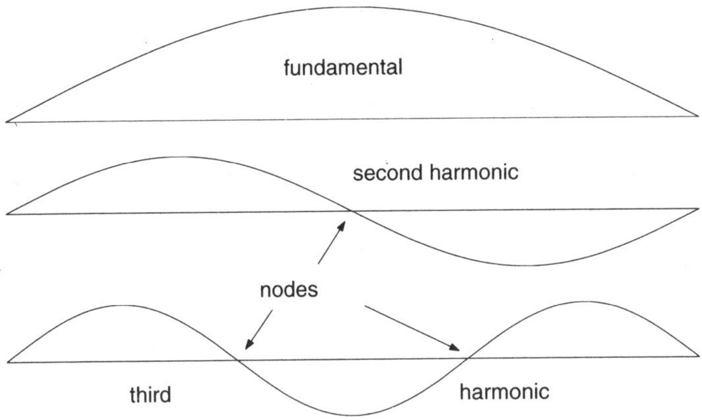

Figure 2.2. *The string's first three natural modes: C, C, G*

Finally, the coefficients c_n are determined by considering the *initial conditions*: the position $u_0(x)$ and the speed $v_0(x)$ of the string at the time $t = 0$. Because the string is bound at both ends, we have $u_0(0) = u_0(L) = 0$. If we extend $u_0(x)$ and $v_0(x)$ *periodically* over the interval $[-L, 0]$, we get two $2L$-periodic functions. These functions can be decomposed in a unique way as series of sines that coincide with $u_0(x)$ and $v_0(x)$ in the interval $[0, L]$:

$$u_0(x) = \sum_{n \geq 1} u_n \sin(n\pi x/L), \tag{2.5}$$

$$v_0(x) = \sum_{n \geq 1} v_n \sin(n\pi x/L). \tag{2.6}$$

The term-by term comparison of these two equations with (2.4) and its derivative with respect to t (the speed), knowing that $k_n = n\pi/L = 2\pi f_n/c$ and $\sin(2\pi k_{-n}x) = -\sin(2\pi k_n x)$, leads for each $n \geq 1$ to the system

$$c_n - c_{-n} = u_n,$$

$$cni\pi c_n/L + cni\pi c_{-n}/L = v_n,$$

with the determinant $2cni\pi/L \neq 0$, which completely defines the values of c_{-n} and c_n.

2.1.2. *Beats, chords and consonance*

We will now describe the use of beats to tune an instrument, in this case a piano. This process can of course be applied to other instruments, such as the harpsichord, the harmonium, the accordion and the organ, but also to musicians themselves inside a musical ensemble. For now we will ignore inharmonicity, but it will be described in section 2.5.1.1 and taken into account in section 3.3.

A piano is built with three strings for each note, tuned in *unison*, that is at the same frequency (the lower notes only have one or two strings). According to (2.1) and (2.3), a string's fundamental is given by Taylor's law:

$$f_1 = \frac{1}{2L}\sqrt{\frac{T}{\mu}}.$$

Because the length L and the mass per unit length μ are determined by the manufacturer, tuning is achieved by modifying the tension T. Note that the total tension on all of the strings of a contemporary concert piano exceeds twenty tons. The tension for each string is adjusted by turning pins to achieve the desired effect. A tuning fork serves as a reference for tuning (A4, which can vary from 440 Hz to 444 Hz, see section 3.4), and from there the tuner makes comparisons by listening for beats, an effect we will now analyze.

We will start by considering two strings slightly out of unison, yielding two fundamentals $s_1(t)$ and $s_2(t)$ with close frequencies $f_1 < f_2$.

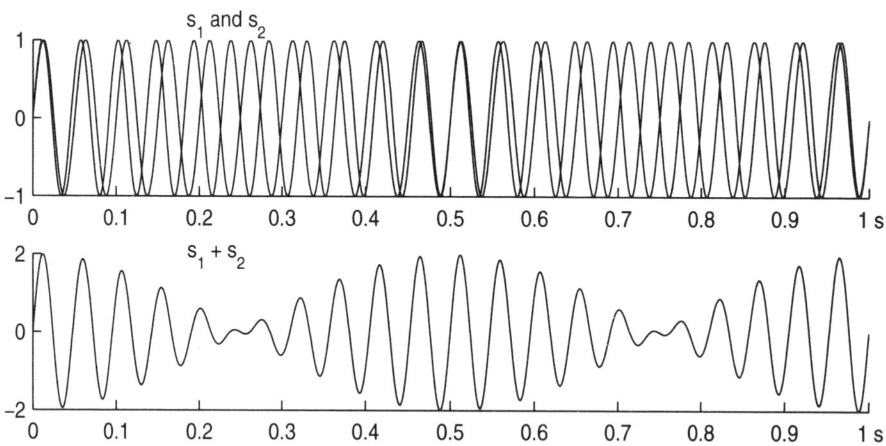

Figure 2.3. *Two close frequencies (top) and their superposition (the sum of the signals, bottom) producing a* beat

The graph on top of Figure 2.3 shows the two fundamentals when

$$s_1(t) = \sin(20 \times 2\pi t), \quad s_2(t) = \sin(22 \times 2\pi t),$$

where the frequencies were chosen for purposes of clarity in the plot. The bottom graph shows their sum $s_1 + s_2$. When the two sines are in phase, they strengthen each other ($t = 0$, $t = 0.5$ and $t = 1$). When they are out of phase, at times $t = 0.25$ and $t = 0.75$, they cancel each other out. This leads to a beat: the sound is amplified and attenuated periodically, with the frequency $f_2 - f_1$ (in this case 2 Hz). This phenomenon can also be inferred from

$$\sin(2\pi f_1 t) + \sin(2\pi f_2 t) = 2\sin(2\pi \frac{f_1 + f_2}{2} t)\cos(2\pi \frac{f_1 - f_2}{2} t).$$

The resulting sound has the intermediate frequency $(f_1 + f_2)/2$, *amplitude modulated* by $\cos(2\pi t(f_1 - f_2)/2)$. The maximum amplitude is reached at times such that $\cos(2\pi t(f_1 - f_2)/2) = \pm 1$, hence the beat frequency is equal to $2|f_1 - f_2|/2 = f_2 - f_1$.

Both strings can be adjusted to the same frequency *simply by eliminating this beat* (see also section 2.6.5). The tuner or the musician does not measure the *frequency of an individual string*, instead he measures *the frequency* (also called the *speed*) of the beat produced by two different strings. This technique is used because a difference of half a hertz for two strings played *consecutively* is completely undetectable by even the sharpest sense of hearing, whereas a beat of half a hertz for strings played *simultaneously* is perfectly detectable by anybody with a bit of practice. The same process can be used to tune intervals other than unison, the difference being that, instead of beats between fundamentals, the tuner has to listen for beats between harmonics of different levels depending on the note tuned. Let us examine the three cases illustrated by Figure 2.4.

– **Tuning the octave**. Let us assume that F3 at 174.6 Hz (in equal temperament) is tuned. The F4 should be tuned at $2 \times 174.6 = 349.2$ Hz. If we assume that it is tuned slightly too high, for example at 351.2 Hz, then the second harmonic of F3, the frequency of which is 349.2 Hz, will produce a beat of $351.2 - 349.2 = 2$ Hz with the fundamental of F4. Tuning will then be achieved by progressively lowering the frequency of F4 until the beat disappears.

– **Tuning the fifth**. We will now consider C4, theoretically with a fundamental of $3 \times 174.6/2$ Hz. Its second harmonic has the frequency 3×174.6 Hz, and so does the third harmonic of F3. The fifth can therefore be tuned by listening to the beat between these two harmonics, until it has almost disappeared.

– **Tuning the major third**. In this case, instead of tuning by suppressing a beat, *the speed is adjusted*. F3 produces a fifth harmonic with frequency 5×174.6 Hz $=$ 873 Hz, whereas C3 produces a fourth harmonic with frequency 4×220 Hz $= 880$ Hz. Hence these two harmonics combined lead to a beat of 7 Hz, detectable with a little effort. The speed of this beat depends of course on the pitch of the third. For example, it will be twice as fast one octave higher. Note that if we reduce the major third (in this case with F3 with frequency 880 Hz$/5 = 176$ Hz), we get a third with no beat, found for example in Zarlino's untempered scale [COR 82, JUN 79]. We will discuss temperament further in Chapter 3.

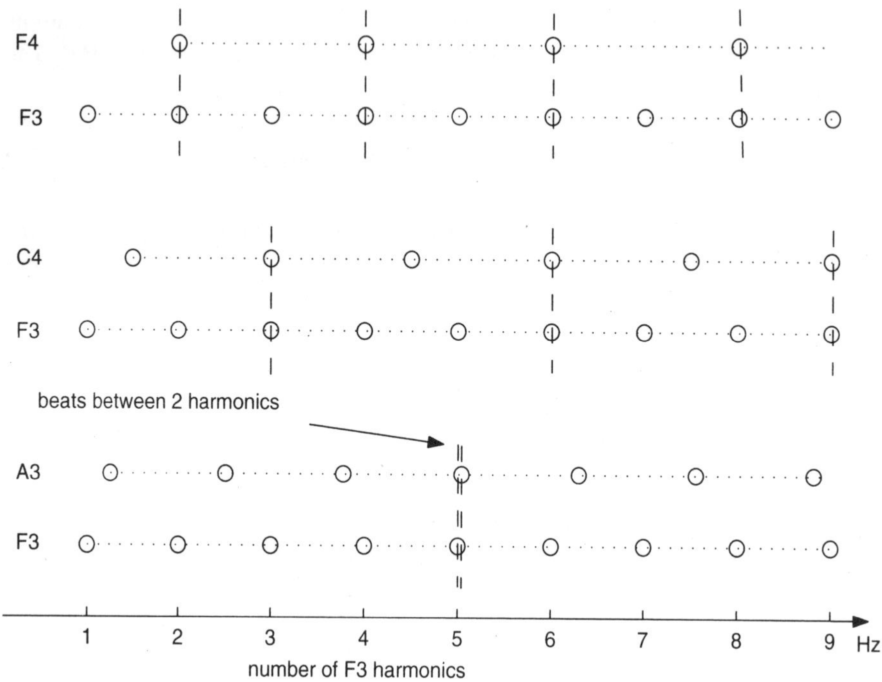

Figure 2.4. *Coincidence of harmonics for the octave and the fifth, slight shift for the equal-tempered major third F-A at harmonics 5 and 4 respectively (the corresponding harmonics' numbers for F3 are shown on the frequency axis)*

If we define, as Helmholtz did in the 19th century [FIC 96, HEL 68], the degree of *consonance* as a description of the extent to which the harmonics of two notes played simultaneously coincide, or instead produce a beat, then the octave is the most consonant interval, followed directly by the fifth, hence the importance of the latter in scale design, which we will study in Chapter 3.

2.2. Bars

Let us set aside strings and now consider the case of a rod or of a bar, with a circular or rectangular section. This is the vibration source for many instruments, such as the accordion, the xylophone (Greek for 'sound of wood'), the vibraphone (a xylophone with metal bars (!), with the addition of tubes that work as resonators and a rotating valve device to modify the amplitude periodically), the celesta (rods struck by a hammer), the Fender Rhodes piano (likewise), music boxes, etc., and in wind instruments, the reed itself!

As was the case with strings, there are several possible types of vibrations, and we will focus on transverse waves in an xOy plane, where the central axis of the bar has its ends at points $(0,0)$ and $(0,L)$. The position of the axis at a time t is given by the equation $y = u(x,t)$. The mechanics model is more complex than with strings, and we will admit that the movement of the bar is governed by the equation

$$\partial_{t^2} u(x,t) = -g^2 c_L^2 \partial_{x^4} u(x,t),$$

where g is the *gyration radius* which depends on the shape of the bar's section[3], $c_L = \sqrt{E/\rho}$ is the propagation speed of the longitudinal waves through the bar, E is the elastic modulus of the material and ρ its density. The harmonic solutions are still of the form

$$u(x,t) = \varphi(x)\exp(2i\pi f t),$$

but $\varphi(x)$ is now a solution to a fourth-order differential equation:

$$\varphi^{(4)}(x) = K^4\varphi(x), \quad K > 0,\ K^4 = \left(\frac{2\pi f}{g c_L}\right)^2 \tag{2.7}$$

The general solution to this homogenous equation can be written as

$$\varphi(x) = \alpha\exp(Kx) + \beta\exp(-Kx) + \gamma\exp(iKx) + \delta\exp(-iKx), \tag{2.8}$$

where the constants are determined from the boundary conditions. We will examine two common types of boundary conditions in music instruments.

2.2.1. *Bar fixed at both ends*

The typical example of the bar fixed at both ends is provided by the xylophone. We will assume a simplified model where the bar is fixed exactly at its ends (which is not quite realistic, but let's move on), and where the function $\varphi(x)$, as well as the second derivatives in the absence of flexion (see Figure 2.5), are equal to 0 in 0 and L. This leads, at point $x = 0$, to:

$$\begin{cases} \alpha + \beta + \gamma + \delta = 0, \\ \alpha + \beta - \gamma - \delta = 0, \end{cases}$$

hence $\alpha = -\beta$ and $\gamma = -\delta$, and therefore[4]

$$\varphi(x)/2 = \alpha\,\mathrm{sh}(Kx) + i\gamma\sin(Kx).$$

3. $g = e\sqrt{12}$ for a bar with thickness e, $g = r/2$ for a cylinder with radius r.

4. As a reminder, $\exp(ix) + \exp(-ix) = 2\cos(x)$ and $\exp(ix) - \exp(-ix) = 2i\sin(x)$, and $\exp(x) + \exp(-x) = 2\,\mathrm{ch}\,(x)$ and $\exp(x) - \exp(-x) = 2\,\mathrm{sh}\,(x)$.

At point $x = L$, the two other boundary conditions lead to

$$\begin{cases} \alpha\,\mathrm{sh}(KL) + i\gamma \sin(KL) = 0, \\ \alpha\,\mathrm{sh}(KL) - i\gamma \sin(KL) = 0. \end{cases} \tag{2.9}$$

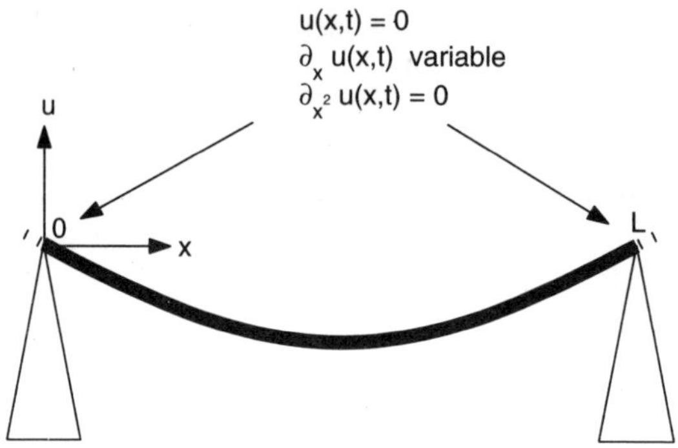

Figure 2.5. *Fixed bar. At both ends, the position is constant, the slope varies with time and the curvature is equal to* 0

System (2.9) with unknowns α and γ has non-zero solutions if and only if $\sin(KL) = 0$, that is to say, if $K = n\pi/L$, with $n \geq 1$, since $K > 0$. In these cases, we have $\alpha = 0$ and, because of (2.7), the harmonic solutions, or natural modes, are of the form

$$u(x,t) = a \sin(K_n x) \exp(\pm 2i\pi f_n t) \tag{2.10}$$

with $K_n = n\pi/L$, and with the natural frequency $f_n = gc_L K_n^2/2\pi$, or

$$f_n = n^2 \frac{gc_L\pi}{2L^2}. \tag{2.11}$$

Notice that, just as for the string, the natural modes are standing waves and share the same form. On the other hand, the natural frequencies f_n now follow a *quadratic* progression

$$1, 4, 9, 16, \ldots, n^2, \ldots$$

as opposed to the *arithmetic* progression of the string's natural frequencies. A similar comment regarding length: the natural frequencies have become inversely proportional to the square of the length. And the last difference: mode (2.10) can be seen as the superposition of two traveling waves $a[\exp(i(2\pi f_n t + K_n x)) - \exp(i(2\pi f_n t - K_n x))]/2i = u(x,t)$ propagating at a speed of $2\pi f_n/K_n = \sqrt{2\pi g c_L f_n}$, now *dependent on the frequency*. This is called *dispersion*, because a traveling wave focused in space will 'spread out', since the high-frequency components travel faster than the low-frequency components.

If we superpose the real parts of the natural modes, in the end we obtain the physical solutions, which can be expressed as

$$u(x,t) = \sum_{n \geq 1} \alpha_n \sin(n\pi x/L)\cos(2\pi f_n t + \theta_n).$$

Because all of the frequencies are integer multiples of the first frequency, the resulting sound is periodic with period $1/f_1$, and lacks certain harmonics (the octave, the fifth above it...), thus contributing to the xylophone's peculiar sonority. The system is solved completely with the help of the initial conditions, just as it was done for the string.

2.2.2. *Bar clamped at one end*

The typical example of a bar clamped at one end and vibrating freely at the other is provided by the music box. In $x = 0$, where the bar is attached, the function $\varphi(x)$ (2.8), as well as its derivative, are equal to 0. At point L, where the bar is not restrained, we will admit that the second and third derivatives are equal to 0. In this case, it is more convenient to express $\varphi(x)$ in the following equivalent form (but with different constants):

$$\varphi(x) = \alpha\,\mathrm{ch}(Kx) + \beta\,\mathrm{sh}(Kx) + \gamma\cos(Kx) + \delta\sin(Kx).$$

The two conditions at point $x = 0$ lead to

$$\begin{cases} \alpha + \gamma - 0, \\ \beta + \delta = 0, \end{cases}$$

hence

$$\varphi(x) = \alpha[\mathrm{ch}(Kx) - \cos(Kx)] + \beta[\mathrm{sh}(Kx) - \sin(Kx)].$$

The two conditions at point $x = L$ can be expressed as

$$\begin{cases} \alpha[\mathrm{ch}(KL) + \cos(KL)] + \beta[\mathrm{sh}(KL) + \sin(KL)] = 0, \\ \alpha[\mathrm{sh}(KL) - \sin(KL)] + \beta[\mathrm{ch}(KL) + \cos(KL)] = 0. \end{cases}$$

This system has non-zero solutions if and only if its determinant is equal to 0:

$$[\mathrm{ch}(KL) + \cos(KL)]^2 - \mathrm{sh}^2(KL) + \sin^2(KL) = 0,$$

which yields after a simplification

$$\frac{1}{\mathrm{ch}(KL)} + \cos(KL) = 0.$$

Let λ_n be the positive solutions, arranged in increasing order, to the transcendental equation

$$\frac{1}{\mathrm{ch}(\lambda_n)} + \cos(\lambda_n) = 0. \tag{2.12}$$

Hence the solutions to (2.7) are obtained for $K_n = \lambda_n / L$ and $f_n = g c_L K_n^2 / 2\pi$, or

$$f_n = \lambda_n^2 \frac{g c_L}{2\pi L^2}, \tag{2.13}$$

and again, as you can see, the natural modes are standing waves. The first four are shown in Figure 2.6.

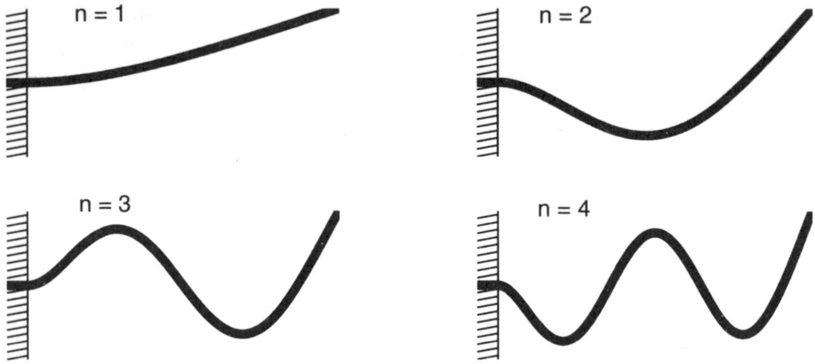

Figure 2.6. *First four natural modes (amplified) of an embedded bar*

Unlike what we saw in the previous example (fixed bar), the natural frequencies f_n are no longer *integer* multiples of the first frequency f_1, or of any other frequency f for that matter. Therefore the frequencies f_n are not harmonics anymore, they are *partials*. Solving (2.12) numerically shows that the values f_n are proportional to the sequence

$$1,\ 6.27,\ 17.55,\ 34.39,\ \ldots$$

Notice that the progression is faster (at the beginning) than for the fixed bar. However, we can infer from (2.12) that, asymptotically, we have $\lambda_n \simeq (n - 1/2)\pi$, where for n high enough

$$f_n \simeq (n - 1/2)^2 \frac{g c_L \pi}{2L^2},$$

a quadratic progression similar to the progression found for the fixed bar (2.11).

As for the physical solution, it is still obtained from summing the harmonic solutions, but usually, this no longer produces a periodic signal. The resulting sound can be considered less 'melodious'. However, we must take into account the resonator or the ear, which can eliminate high partials, as is the case for example with music boxes.

Bars are usually tuned by removing material to modify their thickness, for example near the attached end to reduce the frequency or at the other end to increase it.

2.3. Membranes

Aside from bars, the percussion category also includes instruments comprising an elastic membrane attached to a circular frame: timpani, drums, congas, etc. Let us consider such a membrane with radius R, arranged horizontally, and the height of which is a function $u(x, y, t)$ such that $u(x, y, t) = 0$ if $r = \sqrt{x^2 + y^2} = R$ (*i.e.*, on the frame).

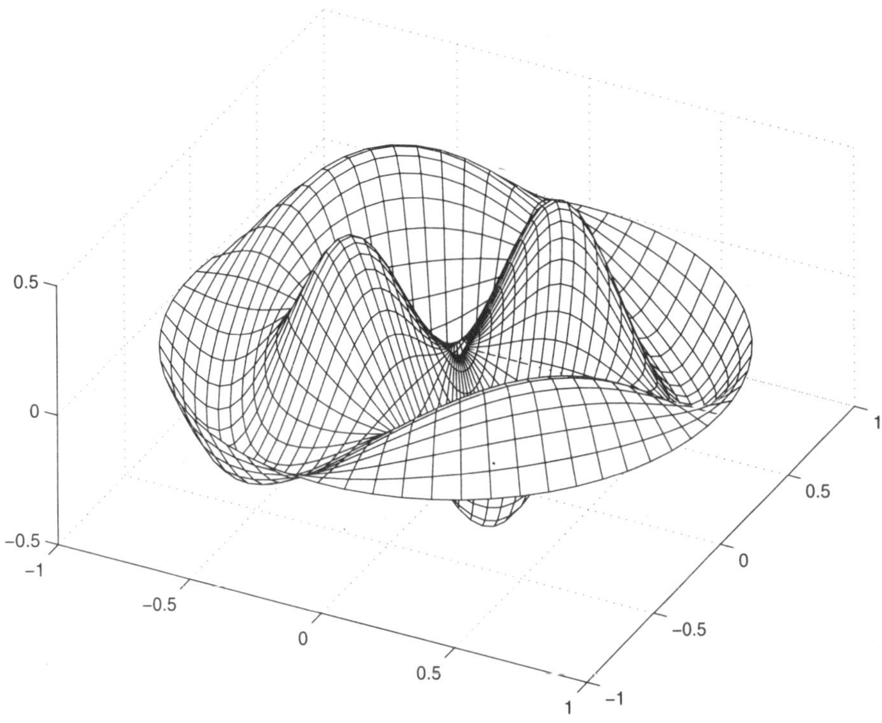

Figure 2.7. *Eighth natural mode of a timpani*

We will assume that the tension T is the same over the entire membrane, and the mass per unit area is denoted by μ. If we follow the same process as with the string, we find that u is a solution to the wave equation (with in this case $\Delta = \partial_{x^2} + \partial_{y^2}$):

$$\mu \partial_{t^2} u(x, y, t) = T \Delta u(x, y, t). \tag{2.14}$$

When substituted in equation (2.14) with $k = 2\pi f / c$ and $c = \sqrt{T/\mu}$, the harmonic solutions, the form of which is $u(x, y, t) = \varphi(x, y) \exp(2i\pi f t)$, lead to the Helmholtz equation:

$$\Delta \varphi(x, y) + k^2 \varphi(x, y) = 0.$$

Because the membrane's edge is circular, it is convenient for purposes of analysis to switch to polar coordinates (r, θ). We are going to search for solutions that can be written in the separated form $\varphi(x, y) = \gamma(r)\sigma(\theta)$, using the formula for the Laplacian in polar coordinates $\Delta = \partial_{r^2} + \partial_r/r + \partial_{\theta^2}/r^2$. After dividing by $\gamma(r)\sigma(\theta)/r^2$, the calculation yields

$$r^2 \frac{\gamma''(r)}{\gamma(r)} + r\frac{\gamma'(r)}{\gamma(r)} + k^2 r^2 = -\frac{\sigma''(\theta)}{\sigma(\theta)}.$$

This expression must be a constant, we will denote it by m^2. On the one hand, we now have $\sigma''(\theta) = -m^2\sigma(\theta)$, with the general solution

$$\sigma(\theta) = \alpha \exp(im\theta) + \beta \exp(-im\theta).$$

Furthermore, because the function σ is 2π-periodic, m must be an integer. On the other hand, we have

$$r^2\gamma''(r) + r\gamma'(r) + (k^2r^2 - m^2)\gamma(r) = 0,$$

which is the *Bessel equation*. Its solutions, when bounded in zero, are the Bessel functions of the first kind of order m, denoted by J_m:

$$\gamma(r) = \alpha J_m(|k|r).$$

We still have to take into account the boundary conditions (on the frame) $\gamma(R) = 0$. This means that we must have $J_m(|k|R) = 0$, therefore, $|k|R$ is a zero of J_m. The zeros of J_m are denoted by z_{mn}.

$$J_m(z_{mn}) = 0.$$

Table 2.1 gives the first approximate values of z_{mn}.

$m\backslash n$	0	1	2	3	4	5
0		2.40	5.52	8.65	11.79	14.93
1	0	3.83	7.02	10.17	13.32	16.47
2	0	5.14	8.42	11.62	14.80	17.96
3	0	6.38	9.76	13.02	16.22	19.41
4	0	7.59	11.06	14.37	17.62	20.83
5	0	8.77	12.34	15.70	18.98	22.22

Table 2.1. *First zeros of the functions J_m*

Hence the harmonic solutions, or natural modes, are of the form

$$u(x, y, t) = J_m(z_{mn}r/R)\left[\alpha \exp(im\theta) + \beta \exp(-im\theta)\right] \exp(\pm 2i\pi f_{mn}t)$$

with the natural frequencies

$$f_{mn} = \frac{z_{mn}c}{2\pi R}. \tag{2.15}$$

These frequencies are partials. Again, the general solutions to the wave equation are obtained by superposing the natural modes.

Unlike the waves we found for strings or bars, these don't have to be standing waves. For $m \geq 1$, we can have rotating waves, clockwise for example such as with

$$u(x, y, t) = \alpha J_m(z_{mn}r/R) \exp(2i\pi f_{mn}t - im\theta),$$

where the angular speed of rotation, in this case, is $2\pi f_{mn}/m$: the value of u is constant at any point moving along a circle centered in $(0, 0)$ and with equation $\theta - 2\pi f_{mn}t/m = \text{constant}$.

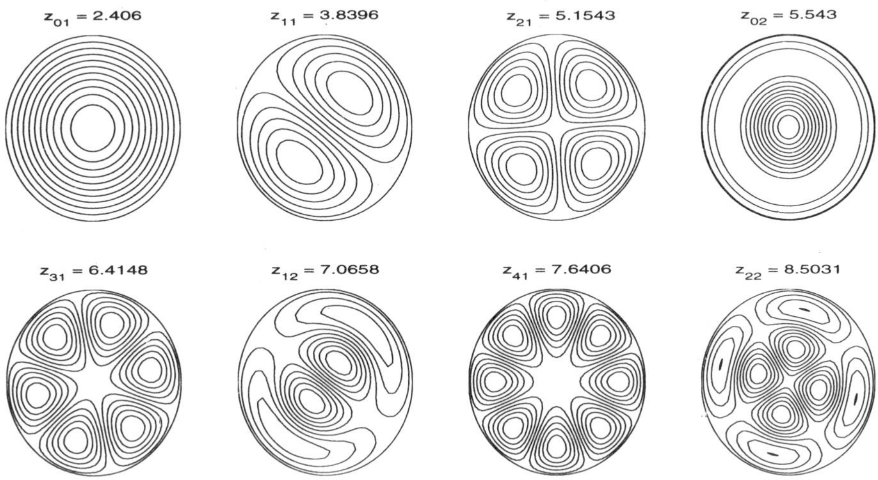

Figure 2.8. *Computed contour lines of the first eight natural modes*

As for the standing natural modes, they are of the form

$$u(x, y, t) = \alpha J_m(z_{mn}r/R) \sin(m(\theta - \theta_0)) \exp(\pm 2i\pi f_{mn}t)..$$

Figure 2.8 shows the contour lines of the first eight modes. The values z_{mn} were computed from the *eigenvalues* of the finite element matrix [LAS 94] associated with the problem, and are close to the values shown in Table 2.1.

2.4. Tubes

Let us now forget about percussions and go back to wind instruments, as we examine their resonator: the column of air contained in the tube. Unlike other instruments we have studied so far, the sound is *sustained*, either directly by the musician blowing into the mouthpiece (woodwind, brass), or by a mechanical blower (organ). We are going to study the case of a *cylindrical* tube with length L (see Figure 2.9), assuming as a hypothesis that a plane wave is traveling through the tube[5] along the tube's axis Ox. Hence the acoustic pressure p_a inside the tube depends only on x and t, and it is simply denoted by $p(x,t)$. The (average) speed of the air particles inside the tube is still denoted by $v(x,t)$. In the simplified model we are describing, the acoustic excitation produced by the mouthpiece is set, and we study the tube's reaction to this excitation. We can then divide the excitations or 'controls' of the acoustic phenomenon into two categories.

– **Pressure control**: in this case, the source of the air vibrations consists of a pressure $p_E(t)$ imposed at the tube's entrance (on the left): $p(0,t) = p_E(t)$ for any t. This model is an approximation of a flute or of a 'flue pipe' on an organ (which works the same way as a flute). We will see that resonance occurs at the natural frequencies of the tube, assumed to be open at both ends.

– **Speed control**: in this case, the speed of the air is imposed at the tube's entrance: $v(0,t) = v_E(t)$ for any t. This model approximately describes reed instruments, such as the clarinet, or a 'reed pipe' on an organ. Resonance then occurs at the natural frequencies of the tube, which is considered open at one end and closed at the other.

As we will see later on, the type of excitation or control we deal with has a major influence on timbre in the case of a cylindrical tube: the presence or the absence of odd harmonics, but also the pitch of the fundamental. Note that this difference decreases, even disappears, in the case of a *conical* tube (oboe, bassoon, saxophone) as Study problems 2.6.10 and 2.6.11 show. Here, the word *control* (pressure control or speed control) refers to the condition imposed at the tube's entrance, not to the one imposed by the musician, who affects not the tube itself but what makes it vibrate: the reed, a flow of air, the lips. In the case of a reed, for example, the musician controls mostly the pressure on the outside part of the reed, and hence, the air flow crossing the mouthpiece – this results for example from the Bernoulli equation $v = \sqrt{2(p_0 - p)/\rho}$, – or in other words the air speed at the tube's entrance. Furthermore, there are other types of controls that are *dual* (see section 2.6.9). For these, the condition is imposed on $\alpha\, p(0,t) + \beta\, v(0,t)$, and the two previous kinds are only particular cases.

At any rate, these are two simplified models, and reality is more complicated. Particularly, tubes do not have to be cylindrical or conical: they can have flared openings at the end such as the bell on brass instruments; the interaction between the source

5. This hypothesis is valid if we assume that the tube's walls are absolutely rigid and sealed. In that case, the speed perpendicular to the wall is equal to zero at the wall, and we infer from the Euler equation that the pressure's normal derivative is equal to zero at the wall: $\nabla p.\mathbf{n} = 0$, where $\mathbf{n}$ is a unit vector perpendicular ('normal') to the wall. This ensures that, for a given t, the pressure is constant at any point of a right section of the tube, and therefore is a plane wave (however other kinds exist).

and the tube can be non-linear, making them more difficult to study, which is why tubes are still the subject of extensive research. Furthermore, we assume that the tube has no other openings than the ones at both ends, which is far from being true: many wind instruments have holes punched on the side to change notes! And to consider that the presence of a hole has roughly the same effect as if the tube had been cut in the same place is a rather crude approximation.

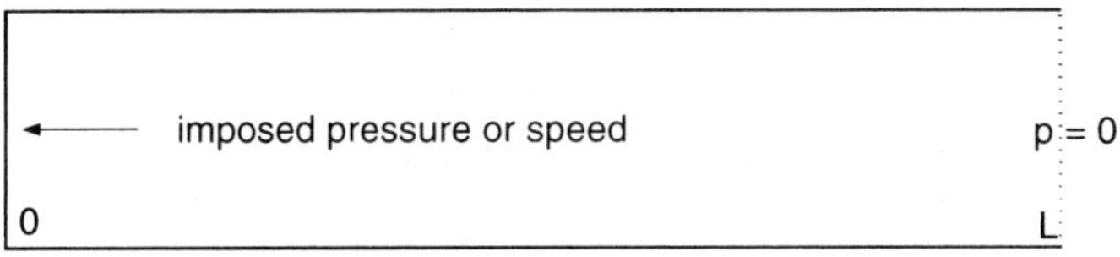

Figure 2.9. *Cylindrical tube with the boundary conditions*

2.4.1. *Pressure control*

The source of the vibration here will be the pressure at the *tube's entrance* $p_E(t)$. Also, it is a reasonable approximation to assume that the acoustic pressure at the tube's exit is zero (the tube is open to the surrounding air). This is not exactly true, and wind instrument manufacturers take this into account by shortening the length of the tube accordingly (by about $0.6\,r$ if r is the radius of the tube) compared to the length in the simplified model. Based on the study of the wave equation conducted in Chapter 1, we get the following system of linear equations:

$$\begin{cases} \partial_{t^2} p(x,t) - c^2 \partial_{x^2} p(x,t) = 0, & \text{inside the tube,} \\ p(0,t) = p_E(t), & \text{at the entrance,} \\ p(L,t) = 0, & \text{at the exit.} \end{cases} \tag{2.16}$$

The source or the excitation $p_E(t)$ is assumed to be periodic, with fundamental f. It can therefore be expanded in a Fourier series

$$p_E(t) = \sum_{n=-\infty}^{+\infty} c_n \exp(2i\pi n f t), \tag{2.17}$$

making it possible to study the produced sound.

2.4.1.1. *Response to a harmonic excitation*

Because system (2.16) is linear, the tube's response will be the sum of the responses to each of the excitations $c_n \exp(2i\pi n f t)$. We will focus on the case where the excitation is harmonic, hence

$$p_E(t) = \exp(2i\pi f t), \tag{2.18}$$

where f can take any value (the coefficient n is omitted). The pressure is then also harmonic, hence of the form

$$p(x,t) = \varphi(x)\exp(2i\pi ft).$$

This expression, substituted in (2.16), shows that $\varphi(x)$ is again a solution to the Helmholtz equation (where $k = 2\pi f/c$), with this time non-homogenous boundary conditions:

$$\begin{cases} \varphi''(x) + k^2\varphi(x) = 0, & \text{inside the tube,} \\ \varphi(0) = 1, & \text{at the entrance,} \\ \varphi(L) = 0, & \text{at the exit.} \end{cases} \tag{2.19}$$

The general solution to the first equation is $\varphi(x) = \alpha\exp(ikx) + \beta\exp(-ikx)$, and the boundary conditions impose

$$\begin{cases} \alpha + \beta = 1, \\ \alpha\exp(ikL) + \beta\exp(-ikL) = 0. \end{cases}$$

This linear system with unknowns α and β has a unique solution if and only if the determinant is different from zero, hence if

$$\sin(kL) \neq 0. \tag{2.20}$$

If this is the case, we get

$$\alpha = i\exp(-ikL)/(2\sin(kL)), \quad \beta = -i\exp(ikL)/(2\sin(kL)),$$

and we have

$$\varphi(x) = \frac{i\exp(-ik(L-x)) - i\exp(ik(L-x))}{2\sin(kL)} = \frac{\sin(k(L-x))}{\sin(kL)}.$$

The pressure inside in the tube subjected to the harmonic excitation (2.18) is therefore

$$p(x,t) = \frac{\sin(k(L-x))}{\sin(kL)}\exp(2i\pi ft). \tag{2.21}$$

Notice that we get a standing wave, just as with strings and bars.

2.4.1.2. *The resonance effect*

What happens for the 'forbidden' values where $\sin(kL) = 0$? In order to answer this question, we must examine more closely what happens at the tube's exit in $x = L$, where the sound is produced before radiating out into the open. Because the pressure at that point is constant and equal to zero, we have to consider its speed. Based on the state equation (1.5), we know that

$$c^2 \rho_0 \partial_x v(x,t) = -\partial_t p(x,t) = -2i\pi f \frac{\sin(k(L-x))}{\sin(kL)} \exp(2i\pi f t).$$

If we integrate this equation with respect to x, we get

$$v(x,t) = \frac{\cos(k(L-x))}{ic\rho_0 \sin(kL)} \exp(2i\pi f t) + g(t). \tag{2.22}$$

The Euler equation (1.3) is used to determine the value of this integration constant (with respect to x), and we then infer that $g'(t) = 0$. Thus the speed is determined, except for a constant, which we will assume equal to zero, hence $g = 0$.

The point of all this is that when we switch over to the physical domain (*i.e.* when we only keep the real parts), for each *entrance pressure* with frequency f

$$p_E(t) = \cos(2\pi f t)$$

there is a corresponding *air speed at the tube's exit*

$$v(L,t) = \frac{1}{c\rho_0 \sin(kL)} \sin(2\pi f t),$$

which increases as $\sin(kL)$ gets closer to 0, and is theoretically *infinite* if $\sin(kL) = 0$, that is, if $k = n\pi/L$. This effect is called *resonance*. A completely similar analysis can be done for other sustained sound instruments, such as a bowed string. In reality, damping occurs because of energy dissipation in the form of heat or radiation, and at the resonance frequency, the exit speed is not actually infinite, merely very large compared to other speeds.

The resonant frequencies associated with these values of k, given by Bernoulli's law

$$f_n = \frac{nc}{2L}, \quad n \geq 1, \tag{2.23}$$

will therefore produce a powerful sound, and will be favored at the expense of others (see Figure 2.10): these are the frequencies that come out when the musician blows into his instrument. Because their progression is proportional to the sequence of integers $1, 2, 3, ...$, we again have harmonics. In fact, the formula is the same as (2.3) which gave us the frequencies for the string (but the c is not the same). Notice that in order for an organ pipe to play the low-pitched C at 32.7 Hz as its fundamental, its length must be $L = 340/(2 \times 32.7) = 5.2$ m $= 17.1$ ft!

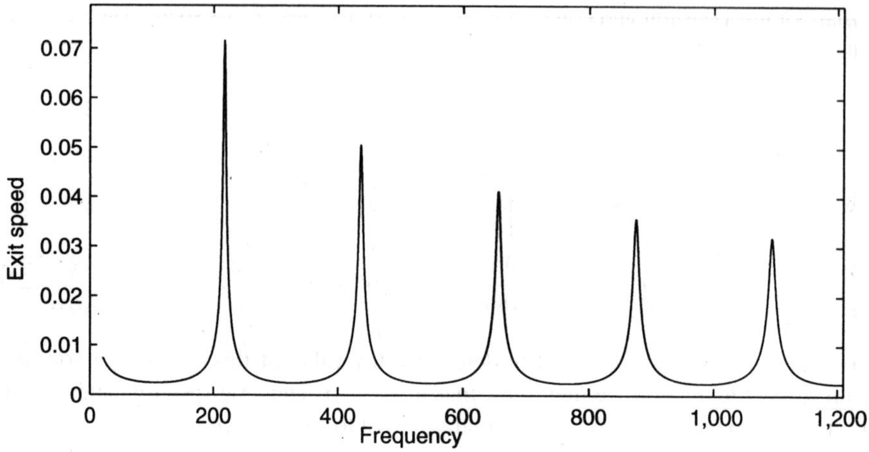

Figure 2.10. *Absolute exit speed for a tube with length 77 cm according to frequency, taking damping into account. The tube resonates at the frequencies where peaks occur*

2.4.1.3. *Natural modes*

If we focus on the limit case where $\sin(kL) = 0$, that is, when $kL = n\pi$, by multiplying equations (2.21) and (2.22) by $\sin(kL)$, we get the resulting functions

$$p(x,t) = \pm\sin(n\pi x/L)\exp(2i\pi ft),$$

$$v(x,t) = \mp\frac{\cos(n\pi x/L)}{ic\rho_0}\exp(2i\pi ft).$$

These functions are still solutions to the wave equation, but no longer satisfy boundary conditions (2.16). Particularly, the pressure is equal to zero at the tube's entrance, whereas at that point, the speed reaches its maximum amplitude. Thus, because the pressure is equal to zero at *both* ends, the natural modes are those of a tube *open at both ends*. These modes occur for frequencies that cause singularities in the pressure control, the natural frequencies (2.23) associated with these natural modes.

2.4.1.4. *The resulting sound*

If we again consider the periodic excitation (2.17) with fundamental f, a powerful sound will only be produced if f (or an integer multiple of f) coincides with one of the tube's natural frequencies (2.23). If for example $f = f_1$, for the listener, the resulting sound will be of the form

$$s(t) = \sum_{n\geq 1} \alpha_n \cos(2n\pi f_1 t + \theta_n),$$

where the amplitudes α_n of the harmonics are proportional to both the coefficients c_n in (2.17) and to the height of the resonance peaks (see Figure 2.10). This is the instrument's *low register*. But if $f = f_2 = 2f_1$, all of the sound's harmonics will be multiplied by two, and the instrument will go one octave higher: this is called *octaving*. It happens in particular when blowing harder in a flute.

2.4.2. *Speed control*

In reed instruments, as we mentioned at the beginning of this section, the vibrations of the column of air are mostly controlled by the speed of the air *at the tube's entrance*. The reed acts as a valve, alternately open or partially closed as the air flows through it, depending on how far the reed is from the mouthpiece. Note that when the musician creates a strong pressure, the air opening will tend to close, the opposite of what occurs with the lips of trumpet player.

2.4.2.1. *Response to a harmonic excitation*

Like pressure, speed obeys to the Helmholtz equation. In the harmonic case, for the entrance speed $v_E(t) = \exp(2i\pi ft)$, the speed's expression is $v(x,t) = \varphi(x)\exp(2i\pi ft)$ with

$$\begin{cases} \varphi''(x) + k^2\varphi(x) = 0, & \text{inside the tube,} \\ \varphi(0) = 1, & \text{at the entrance,} \\ \varphi'(L) = 0, & \text{at the exit.} \end{cases} \tag{2.24}$$

The condition at the tube's exit $\varphi'(L) = 0$, referred to as the Neumann condition, is due to the state equation (1.5) and to the condition $p(L,t) = 0$, which lead to

$$c^2\rho_0\partial_x v(L,t) = -\partial_t p(L,t) = 0,$$

$$c^2\rho_0\varphi'(L)\exp(2i\pi ft) = 0.$$

The general solution to the first equation in (2.24) is $\varphi(x) = \alpha\exp(ikx) + \beta\exp(-ikx)$, and the boundary conditions now impose

$$\begin{cases} \alpha + \beta = 1, \\ \alpha ik\exp(ikL) - \beta ik\exp(-ikL) = 0. \end{cases}$$

This linear system (compare with (2.20)) has a unique solution if and only if $\cos(kL) \neq 0$. If this is the case, the calculation yields $\alpha = \exp(-ikL)/(2\cos(kL))$ and we get

$$\varphi(x) = \frac{\cos(k(L-x))}{\cos(kL)}.$$

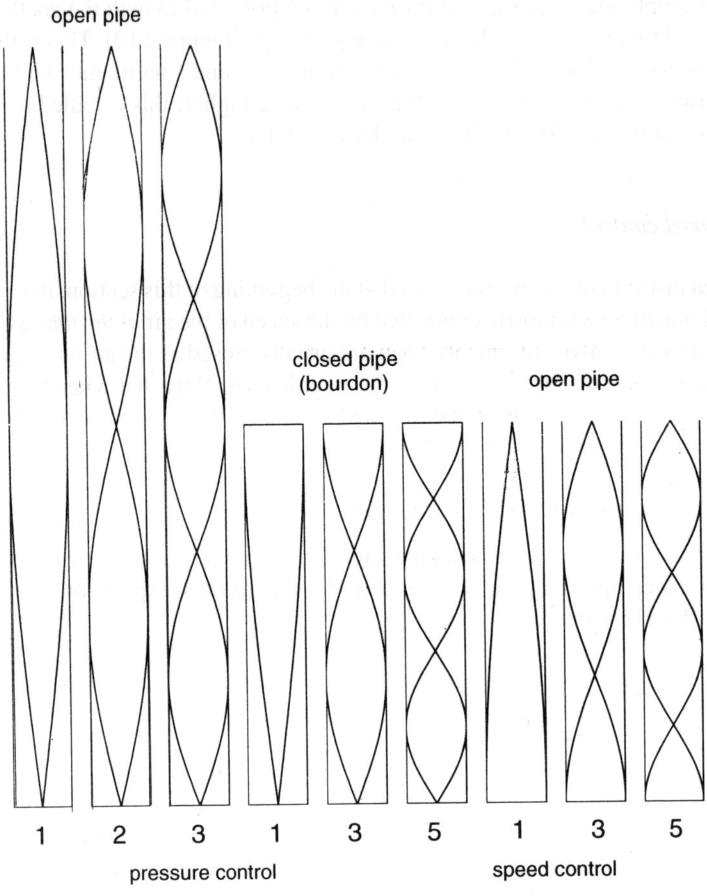

Figure 2.11. *Pressure nodes and antinodes for the first three possible modes, according to the control and to the nature of the exit, open or closed. The numbers indicate the order of the harmonics. Pipes that are twice as short produce the same fundamental, but show no even harmonics.*

Hence the speed inside the tube is

$$v(x,t) = \frac{\cos(k(L-x))}{\cos(kL)} \exp(2i\pi ft), \tag{2.25}$$

and to the *entrance speed* $v(0,t) = \cos(2\pi ft)$ corresponds the *exit speed*

$$v(L,t) = \frac{1}{\cos(kL)} \cos(2\pi ft).$$

2.4.2.2. *Resonance and natural modes*

Notice that the critical values have also changed! The amplified frequencies are no longer the ones for which $\sin(kL) = 0$ (in the case of pressure control), but instead those for which $\cos(2\pi fL/c) = \cos(kL) = 0$, meaning that the natural frequencies are

$$f_n = \frac{(n-1/2)c}{2L}, \quad n \geq 1.$$

We then have two interesting observations to make.

– The fundamental's value is now

$$f_1 = \frac{c}{4L},$$

half of what we observed with pressure control. In this operating mode, the instrument plays one *octave below* – very useful when size is an issue! The same effect actually occurs with a pressure controlled tube, *closed at the other end* (see section 2.6.8). In organs, such pipes are called *bourdons* (see Figure 2.11).

– The sequence of natural frequencies is given by $f_n = (2n-1)f_1$: their progression is now proportional to *odd* integers – this is the price to pay for cutting down on length:

$$1, 3, 5, \ldots, 2n-1, \ldots$$

even harmonics are gone! This absence is actually one of the elements that allow a listener to recognize a reed instrument such as the clarinet, since it is the reason for its 'nasal' sound. This also explains why the clarinet plays low-pitched sounds 'in fifths': when the musician blows harder, he causes the instrument to switch directly from the low register (the *chalumeau* register) to the register one fifth and one octave above (the *bugle* register), instead of octaving, that is, playing one octave higher, as we saw before for the flute.

If we examine the limit case where $\cos(kL) = 0$, hence when $kL = (n + 1/2)\pi$, by multiplying (2.25) by $\cos(kL)$, we obtain a new function

$$v(x,t) = \pm \sin(kx) \exp(2i\pi ft),$$

which is still a solution to the wave equation, but with a speed equal to 0 at the entrance $x = 0$. As for the pressure, it reaches its maximum at the entrance and is always equal to 0 at the exit. Therefore what we have here is the natural mode of a tube *open at one end and closed at the other*.

2.4.2.3. *Comments on phases*

If we superpose the different harmonic modes, we get the resulting sound

$$s(t) = \sum_{n \geq 1} \alpha_n \cos((2n-1)\pi f_1 t + \theta_n),$$

a periodic sound with pitch f_1. Remember that such a sum can be expressed indifferently either with sines or cosines so long as the phases θ_n are included.

Phases can have major effects on the sound's shape. Figure 2.12 shows both sounds

$$s_1(t) = \sum_{n=1}^{10} \frac{1}{2n-1} \sin((2n-1)\pi t),$$

$$s_2(t) = \sum_{n=1}^{10} \frac{1}{2n-1} \cos((2n-1)\pi t).$$

Graphically, the difference seems important, and yet those are exactly the same frequencies. However, listening to these two signals only shows a small difference: it seems the ear is not very sensitive to the phases of frequencies when they are clearly separated from one another. In the case of the clarinet, both of these forms can be observed (among other intermediate ones), depending on the pitch and intensity of the note played [FLE 98].

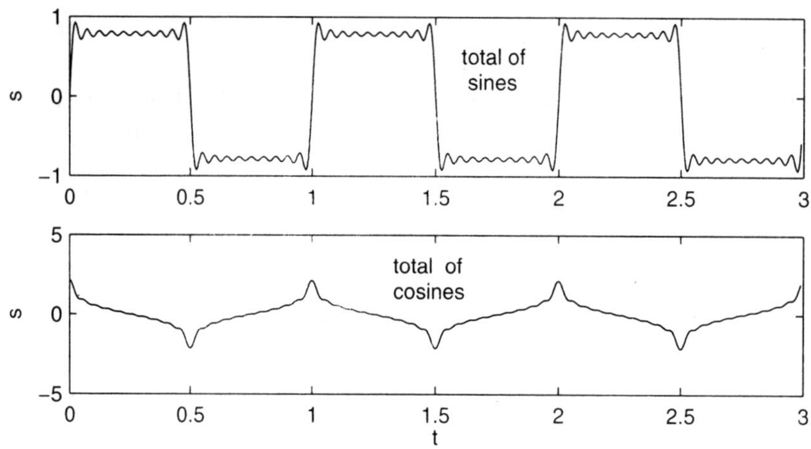

Figure 2.12. *Despite appearances, these two signals are made up of exactly the same frequencies! What makes them different is the phases of the harmonics*

2.4.3. *Tuning*

For string instruments, tuning is done (in addition to the manufacturing process) by adjusting the tube's length near the mouthpiece. For organs, this can be done in several different ways: either by cutting notches in the tube, by moving a sliding ring along the end of the tube (Figure 2.13), also by adjusting the flared-out shape of the end of the tube, or additionally by adjusting the vibrating length of the reed for reed pipes. Since an organ can comprise several thousand pipes (the organ at the Sydney opera has 10,500), this requires a considerable amount of work!

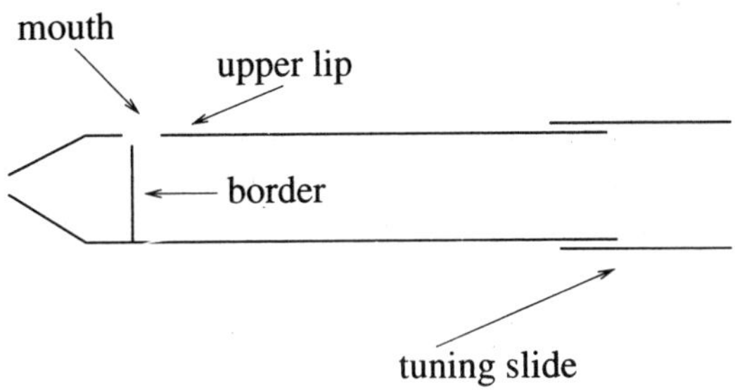

Figure 2.13. *Tuning by adjusting the length*

2.5. Timbre of instruments

Defining the *timbre* of an instrument is no simple task. The literature on the subject is as abundant as it is diverse, with such important works in the 20th century as the *Traité des objets musicaux* [SCH 66] by P. Schaeffer (*A treatise on musical objects*). In this chapter, we will simply write down and describe two characteristics that allow (sometimes not completely) to tell different instruments apart:

– the nature of the sound's spectrum, which depends on the one hand on the vibrator (string, reed, bar, membrane), and on the other on the resonator (sounding board, pipe) which will amplify and 'give color' to the sound produced;

– the sound's envelope, which defines the way a given musical sound is born, lives and dies.

Many other elements have to be considered, such as the vibrato, the intensity of the air flow for wind instruments or the initial impact for percussion instruments, the reverberation of the other strings in a piano, or also the phase shift (Doppler effect, Leslie effect) used for example by jazz musicians by moving around or spinning their instruments. Viola players also share this strange custom, sometimes making their audience seasick! We will discuss some of these aspects in Chapter 6.

2.5.1. *Nature of the spectrum*

Music instruments produce sounds that basically have a discrete spectrum. Hence these sounds can be expressed as

$$s(t) = \sum_{n \geq 1} \alpha_n \cos(2\pi f_n t + \theta_n),$$

an approximation valid at least over a relatively short interval of time. As a result, describing them amounts to saying what frequencies f_n can be found in that representation, and what phases θ_n and amplitudes α_n are associated with these frequencies. These three sets of data already provide a wide variety of timbres.

2.5.1.1. *Harmonics or partials, the piano's inharmonicity*

One of the major characteristics a musician's ear can perfectly determine is whether a sound is composed of harmonics (remember that each f_n is then an integer multiple of f_1 and the sound is periodic) or partials (any other case). This is one of the differences between sustained sound instruments such as bowed strings or wind instruments, and percussion instruments, such as drums or bars. Even among percussion instruments, a musician can easily distinguish the piano's almost periodic sound from the much less periodic sound of a bell, even though these two instruments have similar envelopes, characterized by an impact followed by a sharp decrease.

However, some small size pianos (small upright pianos, baby grands), with shorter strings compensated by a larger diameter[6], produce a slightly acid sound that actually sounds something like a bell. This is due precisely to the fact that the frequencies produced by such a string deviate substantially from the arithmetic progression $f_n = nf_1$ valid for any ideal string, without any stiffness, theoretically 'bendable' at will. The equation governing the movement of a real string is actually a combination of the wave equation for the ideal string and of the equation for the rod fixed at both ends (see section 2.6.6 for a complete study). As a result, the behavior of a real string is a combination of the ideal string and of the rod, for which we saw that the natural frequencies follow an n^2 progression. The result is a progressive shift of the harmonics – which actually become partials – toward the high frequencies: this is called *inharmonicity*, and it increases as the string stiffens, that is to say when the string starts behaving like a rod.

A logarithmic unit is used to measure this inharmonicity, the *cent* or hundredth of a semitone. A tempered semitone means a frequency ratio equal to $2^{1/12} \simeq 1.05946$, and the cent, as a result, means a frequency ratio equal to $2^{1/1\,200} \simeq 1.0005778$. Saying that two frequencies $f_1 < f_2$ are 1 cent apart means that

$$\log_2 f_2 = \log_2(f_1 \times 2^{1/1\,200}) = \log_2 f_1 + \frac{1}{1\,200} = \log_2 f_1 + 1 \text{ cent},$$

6. Taylor's law $f = \sqrt{T/\mu}/(2L)$ tells us that the frequency does not change if the value $L\sqrt{\mu}$ remains constant, where μ is the mass per unit length, which is proportional to the square of the diameter. For example, if the length is divided by 2 and the diameter is multiplied by 2 without the tension being modified, the same note is produced. The problem is that the string's stiffness is increased.

where $\log_2$ is the base 2 logarithm ($\log_2 2^n = n$), which is convenient to use since a one octave interval happens to be a ratio of 2. Piano manufacturers have known for a long time that for a note with theoretical fundamental f, the shift, in cents, of a partial from the corresponding harmonic is roughly proportional to the square of the harmonic's order:

$$\log_2 f_n \simeq \log_2(nf) + \frac{\zeta(f)n^2}{1\,200}.$$

The value of $\zeta(f)$ depends on the note and the instrument. [JUN 79] gives the formula

$$\zeta(f) \simeq \frac{d^2}{L^4 f^2} 3.3\, 10^9$$

where d is the diameter of the string, L its length and f its frequency, thus confirming the study suggested in section 2.6.6. For example, an intermediate frequency on a 107 cm tall piano (42 in) leads to a value of $\zeta(f) \simeq 0.3$. Hence the tenth partial ends up 30 cents above the corresponding harmonic, or a third of a tone! For a grand piano with a string diameter divided by $\sqrt{2}$ but twice as long (and twice the tension to maintain the same frequency), this value is divided by $2 \times 2^4 = 32$, and the shift will only be one cent. Any pianist will tell the difference, even if he doesn't know what causes it.

2.5.1.2. *Richness in higher harmonics*

Another of the timbre's characteristics is the harmonics (or partials) distribution: a sound rich in higher harmonics will be described as 'bright', even 'metallic', whereas a sound poor in higher harmonics will be qualified as 'dull', 'warm', even 'dark'. The first category includes for example the violin, the harpsichord, pianos with very hard hammers. The second includes flutes, some types of organs, pianos with soft felt hammers. Figures 2.14 and 2.15 show the sounds produced by a C flute and by a harpsichord, respectively [COL 98]. The difference in depth for the higher harmonics is striking.

But coming up with categories for instruments is questionable, since even a given instrument, depending on how it is played, particularly the intensity, will change the number of higher harmonics produced: in just about every instrument, the *relative* intensity of the higher harmonics (the ratio $|c_n/c_1|$) increases when a note is played more intensely (see Figure 6.1), which modifies the timbre, a typical *non-linear* behavior of the instrument.

One way of analyzing a sound's harmonics distribution is to observe how fast the Fourier coefficients (the amplitudes of the harmonics) decrease. A mathematical result states that if a periodic function's derivatives up to the order m are square integrable, then its Fourier coefficients c_n are such that

$$\sum_{n=-\infty}^{\infty} |n^m c_n|^2 < \infty, \tag{2.26}$$

and in particular, they decrease faster than $1/n^m$: because the general term of the series tends to 0, we have

$$|c_n| = \frac{\varepsilon(n)}{n^m} \text{ with } \lim_{n \to \infty} \varepsilon(n) = 0.$$

Now let us consider again the example of a struck or plucked string vibrating freely (see (2.4), (2.5), (2.6), as well as section 2.6.2), and notice that the Fourier coefficients are determined by the initial conditions. If the initial conditions are 'not smooth for differentiation' – a string plucked with the jack's quill in a harpsichord, with a fingernail or a pick on a guitar – then the Fourier coefficients will show a relatively slow decrease, hence a sound rich in higher harmonics, and vice versa.

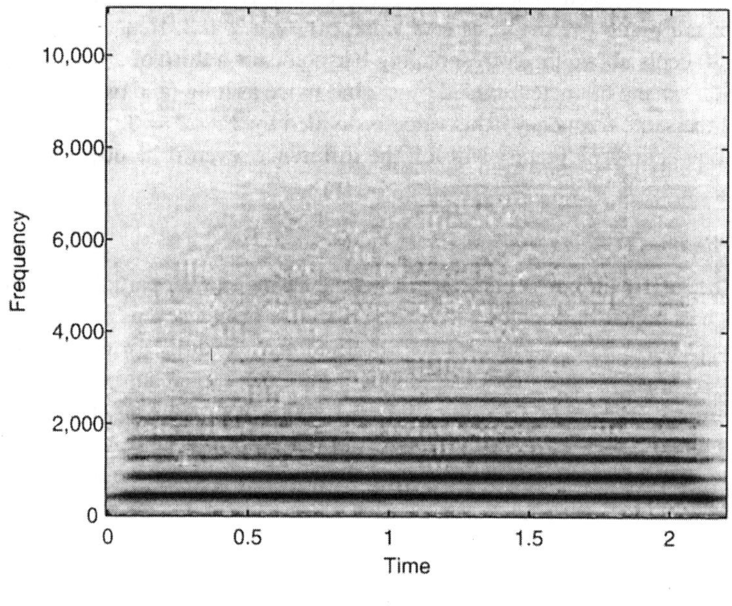

Figure 2.14. *Harmonics of a C flute (G♮-4) with four small keys*

Likewise, with a violin, the bow imposes a jagged motion to the string, referred to as a triangular signal, illustrated in Figure 2.16 (see also section 2.6.3). Every period, the string is first dragged along with the bow until the string's tension overcomes the frictional forces, causing the string to abruptly return to its previous position. Such a signal corresponds to the case $m = 1$ in (2.26), its coefficients are proportional to $1/n^2$, and the limit case of an infinite 'return' slope (discontinuous signal) would lead to coefficients proportional to $1/n$. The decrease is relatively slow, hence, again, a sound rich in higher harmonics.

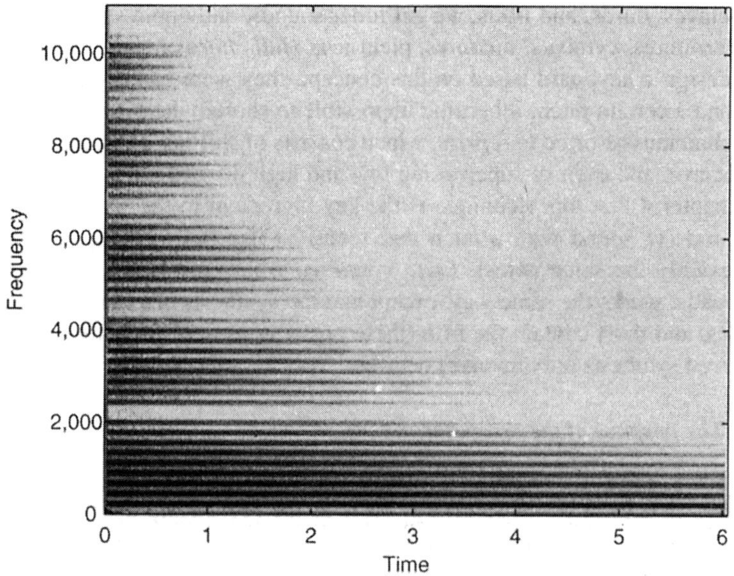

Figure 2.15. *Harpsichord note (A3) plucked at a tenth of its length, which attenuates the harmonics with ranks that are multiples of 10*

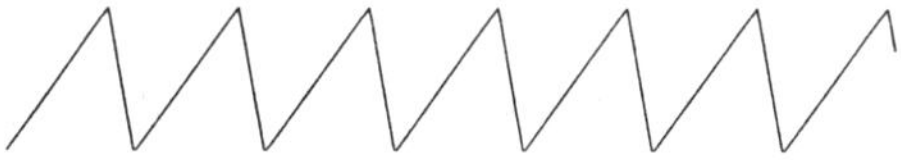

Figure 2.16. *What a saw...*

2.5.1.3. *Different harmonics distributions*

As we have seen already, another unmistakable characteristic of a periodic sound is the virtual absence of even harmonics. This is the case in particular with the clarinet and the bourdon on the organ. Likewise, a large number of different timbres can be obtained by favoring a certain category of harmonics over another. This technique is used extensively by organ manufacturers.

When synthesizing principals[7] [UNI 80], the manufacturers combine *several pipes* to play the same note, for example C2, *corresponding to the harmonics progression*: C2 (principal), C3 (prestant), G3 (fifth), C4 (doublette), E4 (third), G4 (fifth), C5 (sif-flet)... Notice the absence of the seventh harmonic, deemed unaesthetic. By adding

7. Principals are a type of organ stops called flue-stops, of medium 'size' (diameter to length ratio) compared to large-sized stops (flutes and bourdons), which have a rather soft sound, and small-sized stops (gambas and salcionals) designed with the intent of imitating strings.

higher octaves, thirds, and fifths, we get fuller sounds, and organ stops with colorful names: furnitures, cymbals, mixtures, plein-jeux (*full-chorus*). Note that when some tried to design a keyboard based on this concept, they were faced with the problem that beyond a certain pitch, it became impossible to shorten the pipes! Consequently, organ technicians resorted to *reprise*, which consists of shifting down the added notes by one octave, and even of superposing low and high pitched notes! In fact, we will see in Chapter 4 that this technique is the key ingredient to the 'perpetually rising' Shepard tone (a sound with a pitch that seems to rise indefinitely, even though it repeats exactly the same notes). *Flute synthesis*, which involves larger pipes, with softer sounds, works the same way, except that the synthesis is limited to the first six harmonics, and must contain the fifth (these organ stops are called third and cornet). Finally, reed synthesis only involves octaves.

2.5.1.4. *The purpose of the resonator*

Every music instrument uses a resonator to efficiently radiate the energy produced by the musician; even an instrument as basic as the Jew's Harp uses the mouth as a resonator. For string instruments, the resonator is a sounding board made of spruce wood, that receives the string's mechanical energy through the bridge (Figure 2.17). In wind instruments, the resonator is the tube itself. In percussion instruments, it is usually a shell.

The resonator's response, to a first approximation, happens to always be linear: for example the violin's response to two strings played simultaneously is equal to the sum of the responses to the strings played separately. It also turns out that it remains constant with time (see (1.19)): simply put, listening tomorrow to what you played today is the same as waiting a day to listen to what you will play tomorrow. This may seem trivial (or even stupid), and yet, these two hypotheses allow us to state that *the resonator is a filter* (see section 1.5); it amplifies certain frequencies, attenuates others, but never modifies the frequency of a given harmonic. For example, the plot in Figure 2.10 is none other than the transfer function of a wind instrument. The job of an instrument manufacturer can be summed up mainly as 'adjusting' the transfer function to suit the musician's preference. It can sometimes be modified by the musician himself, such as by covering the bell on a trumpet with his or her hand, or with a mute.

Hence, as we saw in the example from Figure 2.10, the amplified frequencies *are the resonator's natural frequencies*. This can sometimes result in unwanted effects, such as what occurs on an instrument with a natural mode that stands out too much. For example, the violin's first resonance, a cavity resonance, is located near the C♯, and is followed just above it by a 'gap' in the response [FLE 98], near the D. Thus, if the frequency of the C♯ is denoted by f_1, the frequency of the D by f_2, and the resonator's transfer function by $H(f)$, we have $|H(f_2)| \ll |H(f_1)|$. If we assume that the signal $e(t)$ provided by the bridge is sinusoidal with frequency f, then the sound produced by the violin will become $H(f)e(t)$ (see section 1.5). If the two notes are played consecutively, the C♯ will sound louder than the D, forcing the violinist to compensate this difference in intensity with his bow, and he will find the instrument uneven. While we are on the subject of the violin's natural modes, Savart observed around 1830, based on measures using the Chladni method[8], that the bottom of a good violin and its sounding board have their first natural frequencies one semitone apart, thus preventing one of these frequencies from being stressed too strongly (these measures were made on Stradivarius and Guarnerius violins that were taken apart [INS 95]!).

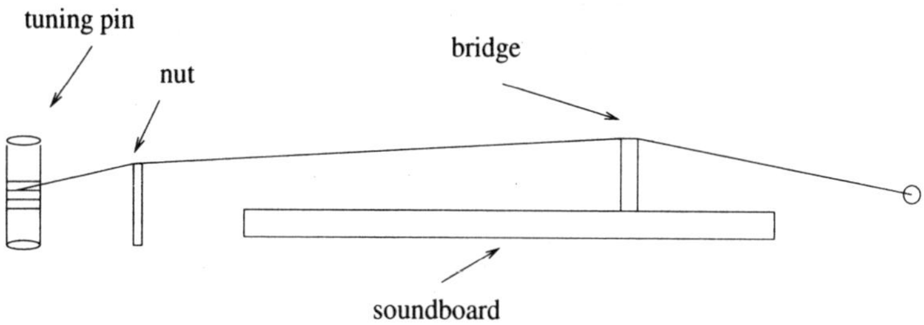

Figure 2.17. *The inseparable string and resonator*

In addition to its role as an amplifier, the resonator always has a direct effect on the spectral distribution of the signal it receives, and therefore plays a fundamental role in the production of the instrument's timbre.

2.5.2. *Envelope of the sound*

Another fundamental characteristic of a sound is its *envelope*, which is basically its time 'packaging', a good example of which was the mountain whistler in Figure 1.20. The envelope defines the way musical sound appears, lives, and disappears. The beginning of the sound, called the *attack point*, has a wide spectrum of frequencies, and plays a crucial role in recognizing the instrument. If this part of the note is removed, many musical sounds become completely impossible to identify, particularly percussion instruments, as P. Schaeffer discovered in 1948. In music synthesis, the attack point can be obtained to some degree by a very rapid increase in the sound intensity at the beginning of the envelope, but this is not really sufficient for a satisfactory result.

In the following example:

$$s(t) = e(t)\sin(60\pi t), \quad 0 \leq t \leq 1, \tag{2.27}$$

8. Which consists of sprinkling small pellets on a horizontal board, and to observe how the pellets arrange themselves when the board is excited by a mechanic vibration with a given frequency f. At the board's natural frequencies, the pellets naturally arrange themselves along the node lines of the associated natural mode.

the sinusoidal signal $\sin(60\pi t)$ is amplitude modulated by a function $e(t) \geq 0$, the envelope of $s(t)$, shown in Figure 2.18. The envelope shown is typical of percussion instruments. It comprises four main periods: a period where the signal rapidly increases, called the attack, which lasts from a few milliseconds up to a few hundredths of a second, a decrease period followed by another, slower decrease period, and a last period where the sound dies out. This is only an example of course, and each period can itself be divided again into several parts. For sustained sound instruments, the envelope can have a very different shape: the attack point is often slower, and the subsequent intensity can remain constant – even increase – for the major part of the note's duration.

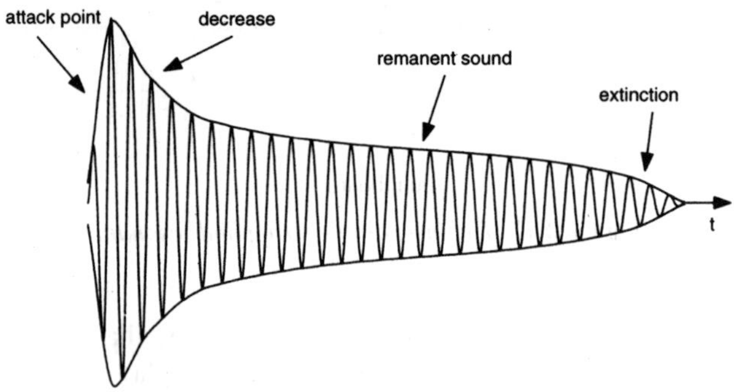

Figure 2.18. *Sinusoidal signal* $\sin 60\pi t$ *amplitude modulated by the envelope* $e(t)$

2.5.2.1. *Calculation of the envelope*

One way of calculating the envelope, and particularly when trying to analyze the sound of an instrument, is based on the *amplitude demodulation* used in radio communications. This is achieved by first passing the signal $s(t)$ into a full-wave rectifier (a diode bridge) delivering a signal $r(t) = |s(t)|$. The rectified signal is then sent through a low-pass filter with impulse response $h(t)$, to 'smooth out' the rapid variations in order to keep nothing but the envelope, and we get

$$e(t) \simeq a(h * |s|)(t),$$

where a is a constant that depends on s. In fact, we can prove the following result.

Consider a signal of the form $s(t) = e(t)v(t)$, *where* $e(t) \geq 0$ *for any* t, *and* $v(t)$ *is* T*-periodic with fundamental* $f_1 = 1/T$. *If we have* $\widehat{e}(f) = 0$ *for any* $|f| > B$ *with* $0 \leq B < f_1/2$, *and if* $h(t)$ *is the ideal low-pass filter (1.26) with cut-off frequency* $f_1/2$, *then we have*

$$e(t) = \frac{1}{c_0}(h * |s|)(t), \quad c_0 = \frac{1}{T}\int_0^T |v(t)|\, dt.$$

Even if the conditions from $\widehat{e}(f)$ and B are not met, but if the variations of $e(t)$ are slow enough compared to $v(t)$, which is the case for instrument sounds, then the difference between $e(t)$ and $(h * |s|)(t)/c_0$ will be small (see Figure 2.19). We will come across the exact same conditions with the Shannon theorem in Chapter 5.

2.5.2.2. *Using several envelopes*

For some instruments, using only one envelope can turn out to be insufficient for an accurate description. It then becomes necessary to resort to a different envelope for each harmonic, and the sound can then be represented as

$$s(t) = \sum_n e_n(t) \sin(2\pi f_n t + \theta_n).$$

To analyze each envelope, the technique from the previous section is applied to each component $e_n(t)\sin(2\pi f_n t+\theta_n)$, which can then be singled-out using the appropriate pass-band filter. Figure 2.19 shows the three envelopes of the first three harmonics (to be precise, they are actually partials) of a piano note, calculated using the method from the previous section, as well as harmonics 1 and 3 singled-out, all of these obtained by applying pass-band filterings to the note. Notice in particular how the third harmonic shows a much sharper drop in intensity than the first.

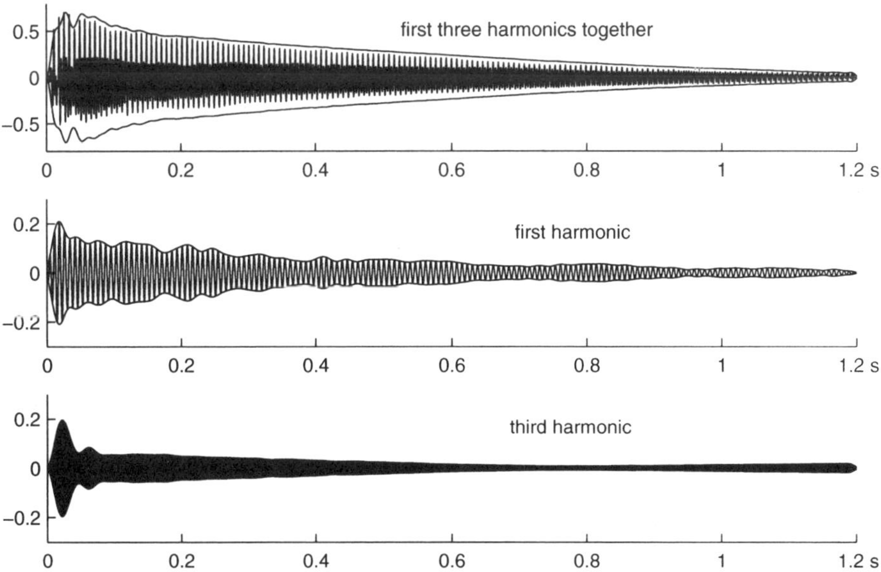

Figure 2.19. *Different harmonic envelopes of C3 (*130.8 *Hz) on a piano*

We end this chapter with two spectrograms illustrating two different behaviors of harmonic envelopes (Figures 2.20 and 2.21). They show that in the sound of a trumpet, higher harmonics are delayed compared to the fundamental (a phenomenon documented by the works of J. C. Risset, see [INS 95]), which accounts in part for the instrument's acoustic signature, whereas in the sound of a piano, the higher harmonics die out more quickly than the lower ones.

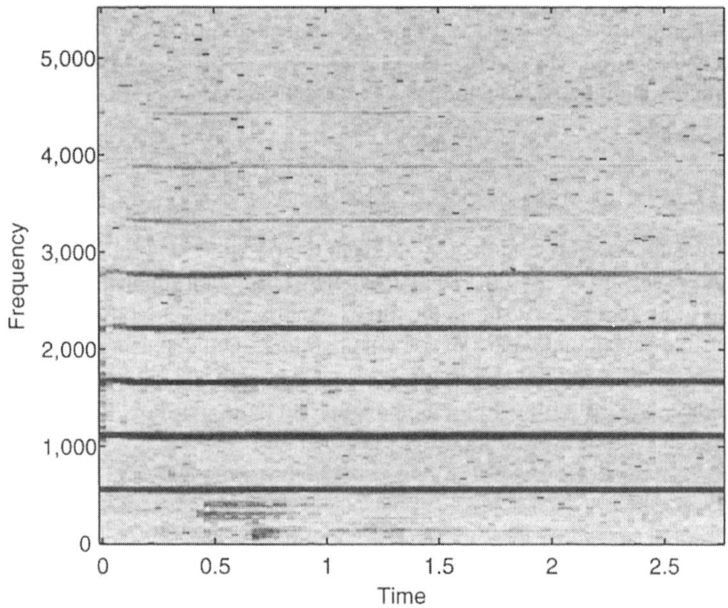

Figure 2.20. *An excerpt of trumpet player Miles Davis in 'The Sorcerer'. He is playing a held note (C♯5), with harmonics coming in later as they get higher*

2.6. Study problems

2.6.1. *Vibrations of a string (general case) (**)*

The movement of a string with length L that is free to vibrate can be determined from the Fourier analysis seen in this chapter. It can also be determined directly using the following method we owe to d'Alembert (1747, [ESC 01]). We already know that this movement can be expressed as

$$u(x,t) = f(x - ct) + g(x + ct).$$

1) Show that the condition $u(0,t) = 0$ for any t implies that $g(y) = -f(-y)$ and therefore

$$u(x,t) = f(x - ct) - f(-x - ct).$$

2) Show that the condition $u(L, t) = 0$ for any t implies that f is $2L$-periodic.

3) f is written as $f(x) = p(x) + q(x)$ where p and q are also $2L$-periodic, with p an even function ($p(x) = p(-x)$) and q an odd function ($q(-x) = -q(x)$). Thus we have

$$u(x,t) = p(x - ct) - p(x + ct) + q(x - ct) + q(x + ct).$$

The initial conditions are given by

$$u(x,0) = u_0(x), \quad \partial_t u(x,0) = v_0(x).$$

Show that

$$q(x) = \frac{1}{2}u_0(x), \quad p(x) = -\frac{1}{2c}V_0(x) + A$$

where $V_0'(x) = v_0(x)$ and A is a constant.

4) These equalities are true for any x if we assume that u_0 and V_0 are extended to be an odd and an even function respectively, both $2L$-periodic. Infer from this result that

$$u(x,t) = \frac{1}{2}\left[u_0(x + ct) + u_0(x - ct)\right] + \frac{1}{2c}\left[V_0(x + ct) - V_0(x - ct)\right]$$

and that this function is T-periodic in time, with $T = 2L/c$.

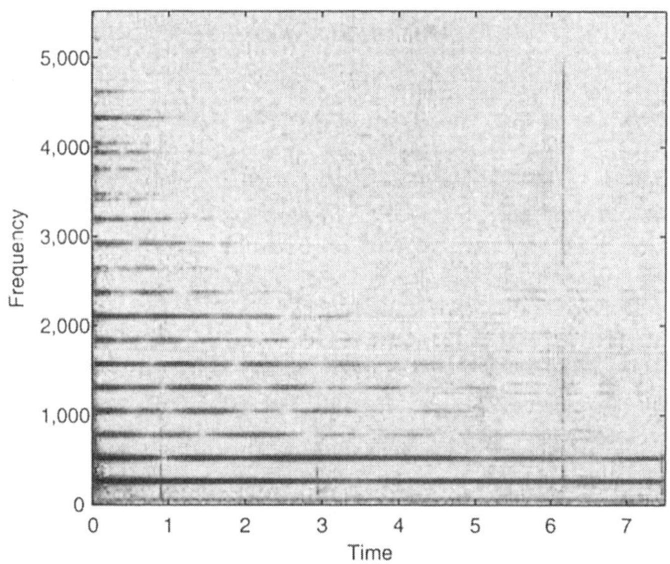

Figure 2.21. *C4 on a piano. The harmonics all start at the same time (even a little earlier for the higher ones), but their duration decreases with their pitch. Notice the beats produced by the three slightly out of tune strings*

2.6.2. *Plucked string (*)*

For a *plucked* string, (guitar, harpsichord), the typical initial conditions are a speed equal to 0, and a piecewise affine position $u_0(x)$ that makes a sharp angle where the string is plucked.

1) Using Study problem 2.6.1, plot the string's position $u(x, t)$ for a few consecutive instants, all taken within the same period.

2) Based on the Fourier analysis, what can be said about the n-th harmonic if the string is plucked at exactly the point with x-coordinate L/n?

2.6.3. *Bow drawn across a string (*)*

In 1877, while he was studying the movement of a string as a bow was *drawn* across it, Helmholtz observed that it underwent a very peculiar deformation, shown in Figure 2.22. The bow has two functions: on the one hand, it is responsible for the shape of the wave, and on the other, its permanent action provides the energy that will be expended by the sounding board. We are going to obtain this movement of the string by assuming that it vibrates freely and without damping, with the following conditions at $t = 0$ for $0 < x < L$ and any α:

$$u(x, 0) = 0, \quad \partial_t u(x, 0) = \alpha(L - x).$$

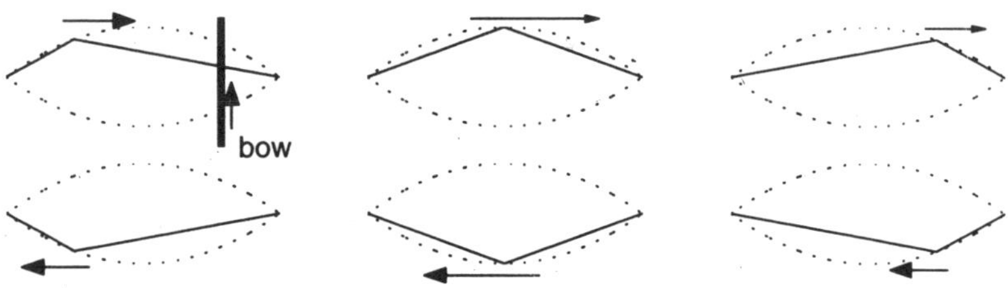

Figure 2.22. *Movement of a bowed string (meant to be read clockwise)*

Using Study problem 2.6.1, which tells us that u is T-periodic in time with $T = 2L/c$, show that over the period $-T/2 \leq t \leq T/2$, we have

$$u(x, t) = \frac{\alpha}{4c}\left[(L - |x - ct|)^2 - (L - |x + ct|)^2\right].$$

COMMENT.– Despite appearances, this function is affine (piecewise) with respect to x and t, because the second degree terms in x^2 and t^2 will cancel out. Hence it accurately describes the movement shown in Figure 2.22.

2.6.4. *String reduced to one degree of freedom (**)*

The most simple model for a string bound at both ends (see Figure 2.23) consists of reducing it to a mobile point M with mass m, connected to the fixed points by two elastic strings with no mass, each exerting on the point M a force T_i, the modulus of which is the string's tension T. In an orthonormal coordinate system xOy, the forces can be expressed as

$$T_1 = -T(\cos\theta, \sin\theta), \quad T_2 = T(\cos\theta, -\sin\theta).$$

The length of the string is L, and point M with coordinates $(L/2, u(t))$ is confined to a vertical line (longitudinal vibrations are discarded).

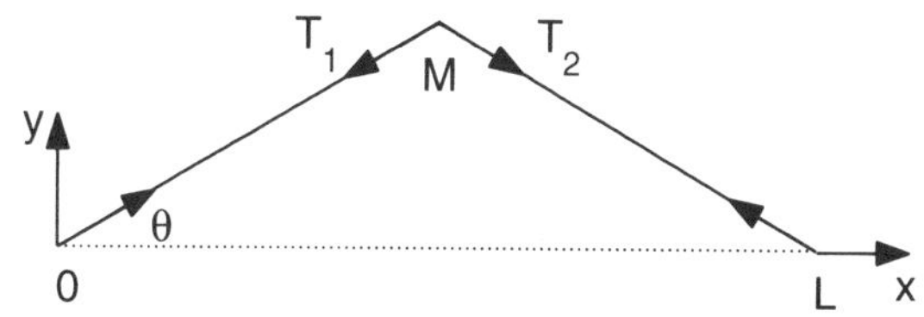

Figure 2.23. *String reduced to a point $M(t)$ with mass m*

1) Using the approximation $\sin\theta \simeq \tan\theta$, show that the force F applied to point M is

$$F = (0, -4uT/L).$$

2) Using Newton's second law of motion, $F = m\gamma$, show that, in the absence of external forces, we have

$$mu''(t) = -Ku(t)$$

where $K = 4T/L$ is the stiffness of the vertical spring equivalent to the two elastic strings.

3) Show that the solutions are sinusoidal vibrations with frequency

$$f = \frac{1}{\pi L}\sqrt{\frac{T}{\mu}}$$

where $\mu = m/L$. Notice that this frequency is smaller than the fundamental of the 'real' string, for which π is replaced with 2. This is due to the fact that, in the simplified model, the entire mass is located in the middle, which increases the string's torque. Lagrange (1759, [ESC 01]) applied this technique to an arbitrary number of masses, then to an infinite number of masses, regularly arranged along the string.

2.6.5. *Coupled string-bridge system and the remanence effect (***)*

Each piano note (except for the low-pitched ones) comprises two or three strings tuned in unison. Typically, the produced sound is composed of two phases: a first phase of quick *damping*, followed by a phase of slower damping, called the *remanent sound.*

Here is a first possible intuitive explanation: at the beginning, the vibrations perpendicular to the sounding board are predominant. They dampen quickly and the parallel vibrations, which dampen more slowly, take over. Remanent sound can actually be caused by a slight mistuning of the strings, as Figure 2.24 shows. It was obtained using a simplified numerical model which we will now describe in detail.

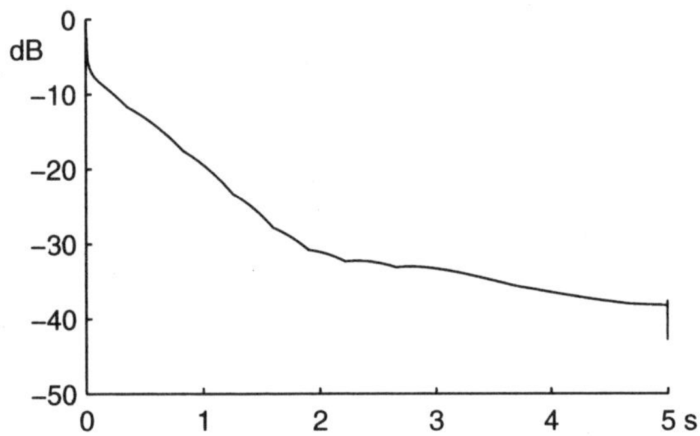

Figure 2.24. *Sound level of a piano. In this case, the remanent sound is caused by a slight mistuning of the strings*

We now consider the same model as before, but this time with strings attached to a bridge that can move. Each string is reduced to a point M_i with mass m and coordinates $(L/2, u_i(t))$, in an orthonormal coordinate system xOy. The bridge is also reduced to a point Q with mass m_c and coordinates $(L, g(t))$. We will assume that it is connected to a fixed point with the same x-coordinate L by means of a shock absorber with stiffness K_c and resistance R. When an external force $b(t)$ is applied (exerted by the strings in what follows), the equation governing the movement of the bridge is therefore

$$m_c g''(t) = -Rg'(t) - K_c g(t) + b(t).$$

1) We begin by examining the interaction of a single string with the bridge (see Figure 2.25). The forces applied to point M are

$$T_1 = -T(\cos(\phi + \theta), \sin(\phi + \theta)), \quad T_2 = T(\cos(\phi - \theta), -\sin(\phi - \theta)),$$

where T is the tension of the string. Using the approximations $\sin\theta \simeq \tan\theta$, $\cos\theta \simeq 1, \ldots$, show that the vertical component of the forces applied to M is equal to $-2T(2u-g)/L$ and use this result to show that

$$mu''(t) = -Ku(t) + Kg(t)/2$$

with $K = 4T/L$. Likewise, show that the vertical component of the forces applied by the strings in Q is equal to $2T(u-g)/L$ and use this result to show that

$$m_c g''(t) = -Rg'(t) - K_c g(t) - Kg(t)/2 + Ku(t)/2.$$

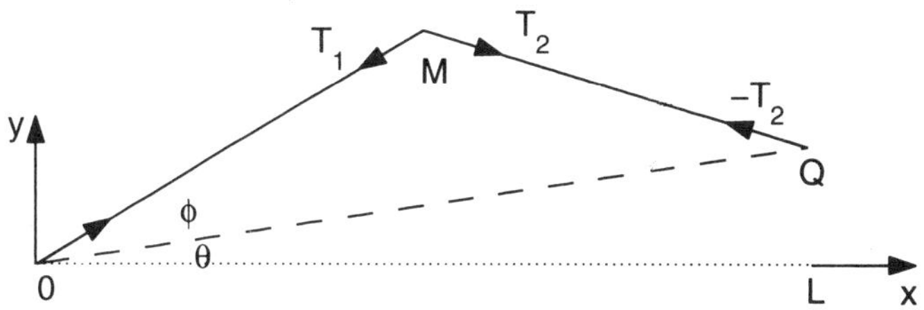

Figure 2.25. *String reduced to a point $M(t)$ with mass m, connected to the bridge, itself reduced to a point $Q(t)$ with mass m_c*

2) Now consider the case of two or three strings $u_i(t)$ subjected to a tension T_i, and let:

$$K_i = \frac{4T_i}{L}, \quad K_s = \sum_{i=1}^{2 \text{ or } 3} K_i.$$

Show that the equations of the complete system become

$$\begin{cases} mu_i''(t) = -K_i u(t) + K_i g(t)/2, \quad i = 1, 2, \ldots \\ m_c g''(t) = -Rg'(t) - (K_c + K_s/2)g(t) + \sum K_i u_i(t)/2. \end{cases} \tag{2.28}$$

3) For the numerical resolution (in this case we will assume that there are 2 strings) let

$$u_i' = v_i, \quad g' = h,$$

$$X = (u_1, u_2, g, v_1, v_2, h).$$

Show that the differential system (2.28) can be expressed in matrix form

$$X'(t) = AX(t)$$

with

$$A = \begin{bmatrix} 0 & 0 & 0 & 1 & 0 & 0 \\ 0 & 0 & 0 & 0 & 1 & 0 \\ 0 & 0 & 0 & 0 & 0 & 1 \\ -K_1/m & 0 & K_1/(2m) & 0 & 0 & 0 \\ 0 & -K_2/m & K_2/(2m) & 0 & 0 & 0 \\ K_1/(2m_c) & K_2/(2m_c) & -(K_c + K_s/2)/m_c & 0 & 0 & -R/m_c \end{bmatrix}.$$

The solution is then

$$X(t) = \exp(tA)X_0$$

where the vector X_0 contains the initial conditions $X(0)$. Numerical resolution consists for example of choosing a time increment τ, then of calculating $M = \exp(\tau A)$, and $X_n = X(n\tau)$ is obtained by induction:

$$X_{n+1} = MX_n.$$

The parameters used in Figure 2.24 are the following:

$$m = 1\,\text{g}, \quad K_i = (2\pi f_i)^2 m\,\text{gs}^{-2} \text{ with } f_1 = 500\,\text{Hz and } f_2 = 500.1\,\text{Hz},$$

$$m_c = 300\,\text{g}, \quad K_c = (2\pi f_c)^2 m_c\,\text{gs}^{-2} \text{ with } f_c = 50\,\text{Hz},$$

$$R = 10\sqrt{K_c m_c}\,\text{gs}^{-1},$$

$$X_0 = \left(0\,\text{m}, 0\,\text{m}, 0\,\text{m}, 1\,\text{ms}^{-1}, .9\,\text{ms}^{-1}, 0\,\text{ms}^{-1}\right).$$

2.6.6. *Calculation of the inharmonicity of a real string (***)*

The equation that governs the movement of a real string, with a non-zero diameter and hence a certain stiffness, is a combination of the equations for an ideal string and a bar. Its expression is as follows:

$$\partial_{t^2} u(x,t) - c^2 \partial_{x^2} u(x,t) + \kappa^2 \partial_{x^4} u(x,t) = 0,$$

where $c^2 = T/\mu$, T being the tension of the string, μ its mass per unit length, and $\kappa^2 = r^2 E/(4\rho)$, where E is the string's Young modulus, r its radius and ρ its density. We will study the harmonic solutions of the form $u(x,t) = \varphi(x)\exp(2i\pi f t)$, which only exist for certain values f_n of the frequency we are going to determine. We will see that these frequencies are no longer harmonics, as they were for the ideal string, but partials.

1) Let $\lambda = \kappa^2/c^2$, show that φ is a solution to the fourth order differential equation

$$\lambda\varphi^{(4)}(x) - \varphi''(x) - k^2\varphi(x) = 0, \tag{2.29}$$

with $k = 2\pi f/c$.

2) The solutions to this equation are linear combinations of exponentials of the form $\exp(Kx)$. Show that $\exp(Kx)$ is a solution if and only if

$$\lambda K^4 - K^2 - k^2 = 0,$$

and infer that the general solution to (2.29) is of the form

$$\varphi(x) = \alpha\exp(K_1x) + \beta\exp(-K_1x) + \gamma\exp(iK_2x) + \delta\exp(-iK_2x) \tag{2.30}$$

with

$$K_1^2 = \frac{1+\sqrt{1+4\lambda k^2}}{2\lambda}, \quad K_2^2 = \frac{-1+\sqrt{1+4\lambda k^2}}{2\lambda}.$$

3) The 4 constants in (2.30) have to meet the conditions imposed by the ends of the string. We will assume that they are as follows:

$$\varphi(0) = 0, \quad \varphi''(0) = 0,$$

$$\varphi(L) = 0, \quad \varphi''(L) = 0.$$

Show that the first two lead to $\alpha = -\beta$ and $\gamma = -\delta$. Hence the solution is of the form

$$\varphi(x)/2 = \alpha\,\mathrm{sh}(K_1x) + i\gamma\sin(K_2x).$$

Now show that the last two conditions lead to non-zero solutions if and only if

$$\sin K_2L = 0$$

(to do this, calculate the determinant of the homogenous system of two equations with two unknowns α and γ).

4) Hence we have the following condition on $K_2 = K_2(n)$, the dependence on n actually affecting $k = k_n = 2\pi f_n/c$:

$$K_2L = n\pi.$$

Let $B = \lambda\pi^2/L^2$, infer from this condition that

$$k_n^2 = \frac{n^2\pi^2}{L^2}(1+Bn^2),$$

and finally show that the partial f_n can be expressed as

$$f_n = nf_1^0\sqrt{1+Bn^2}$$

where $f_1^0 = c/(2L)$ is the string's fundamental in the absence of inharmonicity. Notice that for $r = 0$ (the ideal string), we have $\lambda = 0$ and again find the formula $f_n = nf_1$.

2.6.7. *Coincidence frequency of a wave in a board (***)*

The resonator of many music instruments is made of a wooden board. This board (spruce wood for example) is chosen so as to have a high propagation speed for transverse waves. We will see how this is useful to obtaining a satisfactory sound wave radiation.

To make things simpler, we consider an infinite board, with its median slice located, in its resting position, in the horizontal plane xOy, vibrating vertically at the frequency f, and with its position $z = u(x, y, t)$ on the axis perpendicular to this plane described by a function of the form

$$u(x, y, t) = \cos(kx)\exp(i\omega t),$$

with $\omega = 2\pi f$. The transverse wave equation for a board is written

$$\partial_{t^2}u + \frac{h^2c_L^2}{12}\Delta^2 u = 0, \quad c_L^2 = \frac{E}{\rho(1-\nu^2)}$$

where h is the board's thickness, E, ρ and ν are, respectively, the Young modulus, the density and the Poisson coefficient of the material, assumed to be homogenous and isotropic, c_L is the speed of the longitudinal waves in the board (which we will not be studying here), and $\Delta^2 u = \Delta(\Delta u)$, which in this case amounts to $\Delta^2 u = \partial_{x^4}u$.

1) By assuming $k > 0$, show that

$$k = \sqrt{\frac{\omega\sqrt{12}}{hc_L}},$$

and that the standing transverse wave $u(x, y, t)$ is the superposition of two traveling waves propagating through the board in different directions at the speed (dependent on the frequency)

$$c(\omega) = \frac{\omega}{k} = \sqrt{\frac{\omega h c_L}{\sqrt{12}}}.$$

2) The board's vibrations generate an acoustic wave in the air located in the half-space $z \geq 0$. The components of the air's speed vector are assumed to be 0 except for the z component denoted by $v_z(x, y, z, t)$, which coincides in $z = 0$ with the speed of the board:

$$v_z(x, y, 0, t) = i\omega\cos(kx)\exp(i\omega t).$$

We assume that $v_z(x, y, z, t)$ is of the form

$$v_z(x, y, z, t) = i\omega \cos(kx) \exp(i\omega t - i\kappa z). \tag{2.31}$$

Knowing that v_z is a solution to the wave equation (1.9), show that

$$\kappa^2 = \omega^2 \left(\frac{1}{c^2} - \frac{1}{c(\omega)^2} \right).$$

As a first conclusion, if $c(\omega) < c$, then κ is purely imaginary, and the acoustic wave will decrease exponentially as $\exp(-\alpha z)$ with $0 < \alpha = i\kappa$ (for physical reasons, the exponentially increasing solution is discarded). Furthermore, because this is a standing wave, its intensity over a period T is zero (see section 1.6.4). However, if $c(\omega) > c$, then κ is real, we have a traveling wave, the behavior with respect to z is sinusoidal and the sound 'carries'. The frequency f at which we have the equality $c(\omega) = c$ is called the *coincidence frequency*.

3) We will now assume that $c(\omega) > c$, and focus on the intensity at point x, y, z. The real speed, still denoted by v_z, is (take the real part of (2.31)):

$$v_z(x, y, z, t) = -\omega \cos(kx) \sin(\omega t - \kappa z).$$

Using the state equation $\partial_t p_a = -c^2 \rho_0 (\partial_x v_z + \partial_y v_z + \partial_z v_z)$, show that the acoustic pressure p_a is

$$p_a(x, y, z, t) = c^2 \rho_0 \big[k \sin(kx) \cos(\omega t - \kappa z) - \kappa \cos(kx) \sin(\omega t - \kappa z) \big].$$

Use this result to show that the intensity's value at point x, y, z is

$$I = \frac{c\rho_0 \omega^2 \cos^2(kx)}{2} \sqrt{1 - \frac{c^2}{c(\omega)^2}}.$$

Your conclusion?

2.6.8. *Resonance of the bourdon (**)*

The bourdon on an organ can be considered simply as a tube with pressure control at its entrance in $x = 0$, with the particular feature of being closed at the other end in $x = L$, meaning that at that point:

$$v(L, t) = 0 \text{ for any } t.$$

1) Using Euler's equation, show that the pressure $p(x, t)$ inside the tube meets the boundary condition

$$\partial_x p(L, t) = 0.$$

2) Now go over the study again of the resonance of the pressure controlled tube in section 2.4.1, this time with

$$\begin{cases} \partial_{t^2} p(x,t) - c^2 \partial_{x^2} p(x,t) = 0, & \text{inside the tube,} \\ p(0,t) = p_E(t), & \text{at the entrance,} \\ \partial_x p(L,t) = 0, & \text{at the exit,} \end{cases}$$

and show that the natural frequencies are the same as for the open, speed controlled tube:

$$f_n = \frac{(n - 1/2)c}{2L}, \quad n \geq 1.$$

2.6.9. *Resonance of a cylindrical dual controlled tube (**)*

We now study the case of a harmonic standing wave $p(x,t) = \varphi(x)\exp(2i\pi ft)$ in a cylindrical tube with length L, generated by a *dual control*, that is to say, where the boundary conditions are expressed with real numbers for the values of a and b (this is a particular choice, and not the only possible one):

$$\begin{cases} ap(0,t) + b\partial_x p(0,t) = \exp(2i\pi ft), & \text{at the entrance,} \\ p(L,t) = 0, & \text{at the exit.} \end{cases}$$

1) Remember that $\varphi(x)$ is of the form

$$\varphi(x) = \alpha\exp(ikx) + \beta\exp(-ikx),$$

where $k = 2\pi f/c$. Show that the boundary conditions above impose

$$\begin{cases} \alpha(a + ikb) + \beta(a - ikb) = 1, \\ \alpha\exp(ikL) + \beta\exp(-ikL) = 0, \end{cases}$$

and use this result to show that

$$\alpha = \frac{\exp(-ikL)}{z\exp(-ikL) - \overline{z}\exp(ikL)}, \quad \beta = \frac{-\exp(ikL)}{z\exp(-ikL) - \overline{z}\exp(ikL)},$$

where $z = a + ikb$ and $\overline{z} = a - ikb$ is the conjugate of z.

2) By writing z in the form $z = r\exp(i\theta)$, show that the resonant frequencies, that is, the values of f such that the numbers α and β are not defined, are

$$f_n = \frac{(n\pi + \theta)c}{2\pi L}, \quad n \in \mathbb{Z}.$$

What previous result do we find again for the particular cases $(a, b) = (1, 0)$ and $(a, b) = (0, 1)$?

2.6.10. *Resonance of a conical tube (1) (**)*

A major difference between the clarinet on one hand, and the saxophone, the oboe and the bassoon on the other, is that in the first case the drilling is cylindrical, and in the second it is conical. There lies for the most part the difference in timbre of these instruments, which are all, incidentally, reed instruments (simple or double), and therefore perform to a first approximation with an imposed speed at the tube's entrance. Whereas even harmonics are virtually absent from the sound of a clarinet, we will see that this is far from being the case of the oboe and the saxophone.

Consider a truncated conical tube, the summit of which would be located at the origin, bounded by the sections $r = a$ and $r = b$, with $0 < a < b$, with tube length $L = b - a$ (we are using spherical coordinates with $r = \|x\| = \sqrt{x_1^2 + x_2^2 + x_3^2}$). In harmonic conditions, the acoustic pressure inside the tube can be expressed as

$$p(r,t) = \left[\alpha \frac{\exp(ikr)}{r} + \beta \frac{\exp(-ikr)}{r}\right] \exp(2i\pi ft).$$

1) Let $\mathbf{n}$ be the unit vector, pointing out, perpendicular to the surface located at the tube's entrance ($r = a$). Knowing that the mass conservation equation for an acoustic pressure wave is

$$-\nabla_x p(r,t) = \rho_0 \partial_t v(r,t),$$

show that the speed control $v(r,t).\mathbf{n} = \exp(2i\pi ft)$ imposed at the entrance $r = a$ becomes

$$\partial_r p(a,t) = 2i\pi \rho_0 f \exp(2i\pi ft).$$

2) Assuming that the pressure is equal to 0 at the tube's exit, show that the boundary conditions lead to, in $r = a$ and $r = b$, respectively

$$\begin{cases} \alpha z \exp(ika) + \beta \overline{z} \exp(-ika) = 2i\pi a^2 \rho_0 f, \\ \alpha \exp(ikb) + \beta \exp(-ikb) = 0, \end{cases}$$

where $z = ika - 1$, and infer that

$$\alpha = \frac{2i\pi a^2 \rho_0 f \exp(-ikb)}{z \exp(-ikL) - \overline{z} \exp(ikL)}, \quad \beta = -\alpha \exp(2ikb).$$

3) Show that the resonant frequencies, that is, the values of f for which the numbers α and β are not defined, are solutions to the transcendental equation

$$\tan(2\pi fL/c) + 2\pi fa/c = 0,$$

and use this result to show that the resonant frequencies can be expressed as

$$f_n = \frac{c(-\arctan(2\pi f_n a/c) + n\pi)}{2\pi L}, \qquad n \in \mathbb{Z}.$$

4) Plot the lines with equations $y = \tan(2\pi fL/c)$ and $y = -2\pi fa/c$, which intersect at points with x-coordinates f_n, and based on this result, show that for any $a \ll L$ and n not too high we have

$$f_n \simeq \frac{nc}{2L}.$$

Your conclusion?

2.6.11. *Resonance of a conical tube (2) (**)*

Go over Study problem 2.6.10, this time assuming pressure control, meaning that the conditions imposed at both ends are as follows: $p(a,t) = \exp(2i\pi ft)$ and $p(b,t) = 0$. Solution: the resonant frequencies are exactly the numbers

$$f_n = \frac{nc}{2L}.$$

Notice also that the difference in the harmonics distributions caused by the type of control for a cylindrical tube is clearly attenuated for a conical tube. Conical tubes can grow larger (some organ stops) or smaller (other organ stops, baroque recorder).

2.7. Practical computer applications

This second set of practical applications focuses on creating a synthetic sound imitating an acoustic music instrument, as well as on a few experiments on timbre. The last part can only be done if you have completed Study problem 2.6.5.

2.7.1. *Create your synthesizer*

We are going to write a MATLAB function for the purpose of creating sounds by *additive synthesis* (the summing of sines, see also Chapter 6) and the use of an *envelope* to control the intensity in time. Additive synthesis requires the `synthad.m` function, previously used in the practical applications of Chapter 1.

2.7.1.1. *Write your instrument function*

Create a function with the header

$$\texttt{function s} = \texttt{instrument(f1,T,Fe)}$$

Remember that the file's name is then `instrument.m` (you can replace the word `instrument` with whatever word you like). We first create a sound of the form

$$s(t) = \sum_{k=1}^{np} \alpha_k \sin(2\pi f_1 h_k t), \quad 0 \leq t \leq T.$$

Because the instrument we are dealing with is programmed, the number of harmonics or partials np, their amplitudes α_k and their 'normalized' frequencies h_k, are all defined *by you* in the function itself: the frequency of partial number k is then $f_1 h_k$. In theory (due to the 'normalization' effect), we have $h_1 = 1$ ($h_1 = 0.5$ for a bell), and $f_1 h_1 = f_1$ ($f_1/2$ for a bell) is the frequency of the first harmonic or partial. Once these values are defined, you can use the `synthad` function inside the `instrument` function itself.

A few suggestions regarding the choice of the harmonics or partials distribution: the example of the bell in Chapter 1, the two types of bars, the two types of tubes or the membrane in this chapter. All you have to do now is test your function, for example by executing the following commands:

```
Fe = 22050;
s = instrument(220,3,Fe);
sound(s,Fe);
```

2.7.1.2. *Add an envelope*

Edit the `envelop.m` function you copied from the `AM` website (see instructions at the end of the first chapter). To understand what this function does, execute the following commands:

```
t = [0 .1 .4 1]; a = [0 1 .3 0];
env = envelop(t,a,Fe);
plot(env);
```

then start over with different values (and different numbers of values) in the vectors `t` and `a`.

To add an envelope to the sound you created with your `instrument` function, simply add (at the end of the file) the following command lines *after specifying yourself* in the function the numerical values for the vectors `t` and `a`, while making sure that the first element in `t` is 0 and that the last one is the duration `T`:

```
env = envelop(t,a,Fe);
s = env.*s;
```

Now test your function again.

2.7.1.3. *And play your instrument*

Edit the `play.m` file (copied from the `AM` website) and read it. You are going to make a few modifications. This program calculates the 13 notes of a chromatic scale, starting with the low-pitched frequency `f0` of your choice. You also have to specify in this file the instrument (the function's name) you wish to play. Once this is done, run the `play` command and play!

COMMENT.– In the `windows` system, you will find an application called 'Vienna' that can be used to create 'soundfonts' from samples, and then allows you to play them, either on a virtual keyboard on the screen, or an actual keyboard connected to the computer by a MIDI cable. Such tools also exist in the `linux` system.

2.7.2. *Modify the timbre of your instrument*

Now that your 'synthesizer' works, you can try playing around with the timbre of your instrument by changing the weights of the harmonics or partials in your `instrument.m` file, and listen to the resulting effect with the `play` program. For example, what would happen if we were to remove the first harmonic? If we modified the shape of the envelope? Or the relative weights of the harmonics? If we added some inharmonicity?

2.7.3. *Remanent sound*

The notations and data are the same as in Study problem 2.6.5. Create a command file called `unison.m`, that you will use to program the necessary computations.

The following is a description of the two string model. It relies essentially on programming a loop to compute $X_{n+1} = MX_n$ from $X_n = X(n\tau)$ and $M = \exp(\tau A)$. To do this, the data has to be initialized, particularly X_0 and A (you can use the ones mentioned in Study problem 2.6.5, but it might be more fun to try to find on your own the parameter values that produce the same remanence effect).

Initialization of X_0. We are dealing here with a column vector, the command can be written for example

```
X = [0 0 0 1 .9 0]';
```

(notice the *prime* used for transposing, in this case from a line to a column).

Initialization of A. Follow this example (but the A you will use will have more lines and columns): to initialize the matrix

$$A = \begin{bmatrix} 1 & 2 & 3 \\ 0 & 1 & 1 \\ 2 & 2 & 0 \end{bmatrix},$$

the command is written
`A = [1 2 3; 0 1 1; 2 2 0];`
(notice the ; to move to the next line).

M is computed with the command
`M = exp(tau*A);`
where `tau` is the time increment chosen for the sound representation.

The sound itself is given by the position of the bridge, the third component of the vector X. If `nt` is the number of time increments, then the initialization of the sound vector and of the computation loop are written:

```
sound = zeros(1,nt);
sound(1) = X(3);
for n = 2:nt
   X = M*X;
   sound(n) = X(3);
end
```

Finally, to listen to and visually observe the produced sound, the commands are:
```
soundsc(sound,1/tau);
plot(20*log10(abs(sound)+1e-3));
```
If everything goes according to plan, you should be able to obtain Figure 2.24.

Chapter 3

Scales and Temperaments

When we studied vibrating bodies in Chapter 2, we observed that they can produce either harmonics (strings and tubes) or partials (bells, membranes, percussions). Man may have been more sensitive to the former than the latter, it would seem, when constructing his musical scales. We can suggest the following explanation: strings (the hunter's bow) and tubes (the first known flutes, crafted from bones, date back to 60,000 BC) produce sounds that last longer than with percussions, and it is easier to perceive the harmonics of the former, depending on the degree of their consonance (see section 2.1.2), than the partials of the latter. It is therefore likely that the strong consonance of the fifth[1], which corresponds to a frequency ratio of $3/2$, became predominant very early on in musical history.

We saw earlier that dividing a string into successive portions with lengths $L/1$, $L/2$, $L/3$, $L/4$, $L/5$... produced a sequence of harmonics with frequencies equal to 1, 2, 3, 4, 5... times the fundamental f. The first of these two sequences is said to be harmonic, the second arithmetic. Note that they would have turned up in the reverse order if, instead of shortening the length of the string, we had multiplied its length by 1, 2, 3, 4, 5 to obtain frequencies equal to $f/1$, $f/2$, $f/3$, $f/4$, $f/5$. The first thirteen harmonics are roughly equivalent to the notes given in the following table. The upper line indicates both the ranks of the harmonics and the ratio of their frequency to the frequency of the fundamental, called the *relative frequency*.

1	2	3	4	5	6	7	8	9	10	11	12	13
C	C	G	C	E	G	B♭-	C	D+	E-	F♯-	G	G♯+

1. Resulting from the fact that the low-pitched note's third harmonic coincides with the high-pitched note's second harmonic.

Give or take a few cents (see section 2.5.1), the resulting notes in musical notation are given in Figure 3.1.

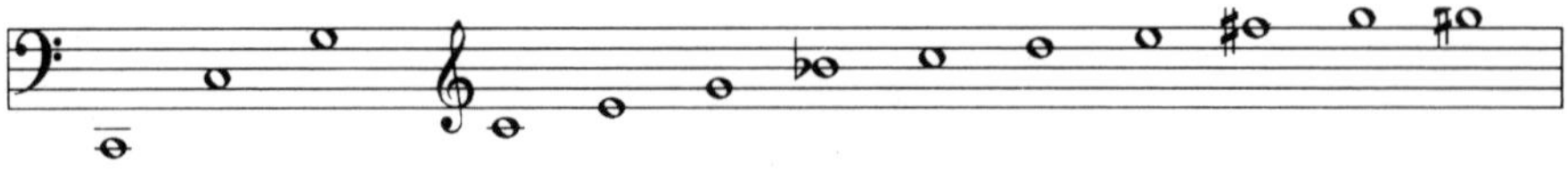

Figure 3.1. *Sequence of the first thirteen harmonics of C2*

If we make all of these notes fit inside one octave, by dividing the frequencies by the appropriate power of 2, and arrange them by order of increasing frequency, we obtain the following frequency ratios:

C	D	E-	F♯-	G	G♯+	B♭-	C
1	$9/8$	$5/4$	$11/8$	$3/2$	$13/8$	$7/4$	2

(3.1)

These notes do not correspond to any classical scale, but you can hear them quite clearly for example when you blow very gently into a C flute, without making the tube resonate.

The countless discussions that took place over the centuries mainly dealt with the first two distinct harmonics of the octave, G and E [BAI 01, FIC 96].

3.1. The Pythagorean scale

In the 6th century BC, Pythagoras was said to give great importance to the arithmetic simplicity of length ratios. Beside the octave ratio 2, the fifth ratio $3/2$ is the 'simplest' of the ratios in Table (3.1) above. Hence Pythagoras emphasized the fifth to construct the diatonic scale named after him, by repeatedly reproducing this interval. The result is the well-known cycle of fifths, F, C, G, D, A, E, B, which leads to the following frequency ratios:

C	D	E	F	G	A	B	C
1	$9/8$	$81/64$	$4/3$	$3/2$	$27/16$	$243/128$	2

Arithmetically speaking, this scale is quite 'elegant', in the sense that it only uses two types of intervals between consecutive notes: the tone, with a ratio of $9/8$, and the semitone, with a ratio of $256/243$. However, aside from the fifth (and the octave of course), it does not coincide with the harmonics. For example, the third with ratio $81/64 \simeq 1.266$ is higher than the corresponding harmonic in Table (3.1) which is equal to $5/4 = 1.25$. The difference, which is difficult to notice inside a melody, becomes noticeable in harmony, that is in a chord, because of the beat effect (see section 2.1.2): for a C with frequency 262 Hz, the harmonic third C-E will produce no beat, whereas the Pythagorean third C-E will produce a perfectly noticeable[2] beat of $|(5 \times 1 - 4 \times 81/64)| \times 262$ Hz $\simeq 16$ Hz. Our ears, which are accustomed to equal temperament, studied later, perceive the Pythagorean third as 'bright' or 'hard'. The beat of the tempered third is much slower: roughly 10 Hz.

Figure 3.2. *Pythagoras's diatonic scale. All the fifths and fourths between consecutive notes 'sound right', in other words the ratios are equal to $3/2$ for the fifths, and to $4/3$ for the fourths (high-pitched/low-pitched frequency ratio)*

3.2. The Zarlino scale

In the 4th century BC, Aristoxenus the Musician, a pupil of Aristotle, recommended in his *Elements of Harmony* 'not to turn one's back on our sensations', and to trust one's ears rather than mathematical reasoning. He therefore suggested to construct the major chord on the basis of harmonics, hence with the frequency ratios 1, $5/4$, $3/2$. This chord produces no beat, and sounds very 'pure'. It was deemed more aesthetic and true to physics by Aristoxenus and his disciples.

Figure 3.3. *The three major chords used to build the Zarlino scale. Harmonic thirds and fifths in each chords. The rest do what they can!*

In the 16th century, Zarlino picked up the idea, and started with the three major chords F-A-C, C-E-G and G-B-D, with the frequency ratios 1, $5/4$, $3/2$, to build his diatonic scale, also called the physicist scale. The frequency ratios are as follows:

C	D	E	F	G	A	B	C
1	$9/8$	$5/4$	$4/3$	$3/2$	$5/3$	$15/8$	2

In this scale, the three major chords with tonic F, C and G are therefore very consonant and produce no beat. But things get ugly when we switch to another tonality. For example, the fifth D-A has a frequency ratio equal to $40/27 \simeq 1.48$, below the 'right' fifth with ratio 1.5. For a D with frequency 294 Hz, the Zarlino fifth D-A produces

2. On the condition of course that the fifth harmonic of the C and the fourth harmonic of the E, which cause the beat, are included in the notes.

a $(3 \times 1 - 2 \times 40/27) \times 294$ Hz $\simeq 11$ Hz beat, which usually sounds dreadful to a musician's ears!

Furthermore, there are now three types of intervals between consecutive notes (instead of two in the Pythagorean scale): the semitone, with the ratio $16/15$, and two types of tones: major tones C-D, F-G and A-B with the ratio $9/8$, and the minor tones B-C and G-A with the ratio $10/9$! The discrepancy between these two types corresponds to a frequency ratio equal to $(9/8)/(10/9) = 81/80 = 1.0125$, an interval called a *syntonic comma* and approximately equal to a ninth of a tone.

All this was not exactly satisfactory, particularly in the context of Western music, which saw the development of polyphony, combined with the desire to be able to transpose in any tonality.

3.3. The tempered scales

J. S. Bach's *Well-Tempered Clavier* was composed at the beginning of the 18th century. It is a collection of preludes and fugues written in all of the twenty-four tonalities, major and minor. As the title of this work suggests, playing these preludes and fugues requires a 'well-tempered' scale (which does not mean equal-tempered), and the Zarlino system, with its very shortened fifths, turns out to be rather inadequate.

But we won't let that stop us. We turn to Pythagoras and proceed with the cycle of fifths that we started, until we have covered the scale's twelve semitones. Note by the way that the notes of the chromatic scale obtained in this manner were already known to the Chinese (the legend attributes them to minister Lin-Louen, 2,600 BC), who used them to transpose the five typical pentatonic scales of Oriental music. Starting with F, the cycle F, C, G, D, A, E, B, F♯, C♯, G♯, D♯, A♯, E♯, will lead to a frequency ratio between the first and last note equal to $(3/2)^{12} \simeq 129.75$. Also, it would be nice if this E♯ could be equal to the *enharmonic* F (the closest F), if only to limit the number of notes on a keyboard. But it just so happens that the enharmonic F of this E♯ corresponds to the ratio $2^7 = 128$. There lies the whole problem:

$$129.75 \simeq (3/2)^{12} \neq 2^7 = 128,$$

making it impossible to have *both* consonant octaves and consonant fifths. It was likewise useless to check this with calculations, since the equality $3^{12} = 128 \times 2^{12}$ is simply impossible: the number on the left is odd whereas the one on the right is even. The gap between the two notes, which corresponds to the ratio $(3/2)^{12}/2^7 \simeq 1.014$, is called a *pythagorean comma*.

3.3.1. *Equal temperament*

A compromise was needed. It was found by Werckmeister around 1695. By setting the equation 12 fifths = 7 octaves, and by favoring the consonance of the octaves, he decided to spread the excess comma among the 12 fifths of the cycle by slightly shortening them. The tempered scales were born, including equal temperament for which all the fifths are equal and correspond to the ratio

$$2^{7/12} \simeq 1.498.$$

Again, the Chinese were a few steps ahead, since equal temperament was invented by prince Chu Tsai-Yu from the Ming dynasty (1368-1643) and officially adopted in 1596, one century before Europe [HON 76].

In equal temperament, the fifth D-A (D at 294 Hz) will produce a beat of

$$(3 \times 1 - 2 \times 2^{7/12}) \times 294\,\text{Hz} \simeq 1\,\text{Hz}, \tag{3.2}$$

which is much more acceptable than the 11 Hz beat produced by the Zarlino fifth D-A. As for the tempered semi-tone, its ratio is

$$2^{1/12} \simeq 1.0595.$$

In this scale, all semitones are equal, and there is no difference between the ♯ and the ♭. On a keyboard, the same key is used to play C♯ and D♭.

3.3.2. *A historical temperament*

In Bach's time, there were actually many temperaments used that were intermediate between Zarlino's temperament and equal temperament. [JUN 79] contains a description of several historical temperaments, including the following one suggested for playing *The Well-Tempered Clavier*. The fact that the tonalities of this temperament are close to C major makes them all the more consonant. Starting with C, we go through a cycle of 4 equal fifths C-G-D-A-E, with a ratio slightly below $3/2$, so as to have a default beat for the fifth C-G and an excess beat for the major third C-E (a ratio higher than $5/4$) both with the same speed: this is the 'well-tempered' clavier. Let x be the ratio of these fifths and f the frequency of the C, this leads to the beats

$$\text{excess beat C-E: } b_1 = |5 - 4x^4/4|f = (x^4 - 5)f,$$

$$\text{default beat C-G: } b_2 = |3 - 2x|f = (3 - 2x)f,$$

and the equal beat speed condition $b_1 = b_2$ leads to

$$x^4 + 2x - 8 = 0,$$

the relevant solution of which is

$$x \simeq 1.496.$$

The fifth B-F♯ is tuned to the same ratio. All the other fifths are tuned to achieve consonance, meaning that they are tuned to obtain the ratio $3/2$. The octaves are 'of course' also tuned to achieve consonance. Quite remarkably, this can be done because we actually do have

$$x^5 \times (3/2)^7 = 128.006 \ldots \simeq 128.$$

Hence this temperament is comprised of major thirds at small, varying distances from the harmonic third, the closest one being that of the 'main' tonality C major, and they increase as their distance to it increases. Thus each tonality has its own sonority, as opposed to what happens with equal temperament.

3.3.3. *Equal temperament with perfect fifth*

Of the fifths and the octaves, Werckmeister chose to make the octaves perfect. But we can ask ourselves this question: why not instead choose to make the fifths perfect? This is precisely what was suggested recently by the piano tuner S. Cordier [COR 82] in what he called *equal temperament with perfect fifth*. Of course, we then end up with slightly widened octaves (see also section 4.3), since the octave ratio must be such that $(3/2)^{12} = x^7$, hence $x \simeq 2.004$. The C4-C5 octave then produces a beat of

$$|2 - 2.004| \times 262 \simeq 1 \text{ Hz}.$$

In other words, the 1 Hz beat of the fifth (3.2), associated with a perfect octave, was passed on to the octave. Widening the octaves corresponds rather well to the practice common to other musicians such as violinists, as well as to the 'distortion' in our perception of pitch, particularly for high-pitched sounds, as we will see in Chapter 4. However, for certain compositions such as for example the sonatas by Beethoven, where double or even triple octaves are frequent, this can turn out to be unpleasant, and so a compromise has to be made between perfect octaves and perfect fifths.

3.3.4. *The practice of tuners*

The theoretical developments we have just discussed, and which are actually mostly relevant to keyboard instruments (the sound of which is fixed), was based on the hypothesis that strings produce harmonics, which is not exactly true, as we saw in section 2.5.1.1. Reality is in fact more complex, and in practice, tuners give the fifths and the octaves a slight beat, as little as possible, default beats for fifths and excess beats for octaves. In other words, they resort to an intermediate temperament between Werckmeister's and Cordier's temperaments, a *maximum consonance equal temperament.*

3.3.5. *The practice of musicians*

Regarding the instrumental practice of musicians who adjust the pitch of the note themselves while they play (violins, wind instruments), equal temperament is far from being the rule. If the fifths are played perfectly, the musicians can usually tell the difference between a C♯ and a D♭, with a clear tendency to shorten the semitones of the leading notes. E. Leipp mentions in his book [LEI 80] that the B, leading note of the C, is sometimes played less than a quarter tone away from the C, and this does not bother the listeners!

As for singing, it is sometimes difficult to determine by analysis the exact pitch of a note. In the excerpt from *Norma* by Maria Callas shown in Figure 1.18, where the tuning fork is set at 441 Hz, the Fourier analysis over 1.8 s of A5 (theoretically 882 Hz) at $t = 7$ s led to Figure 3.4, for which the average pitch seems closer to 900 Hz... But it's Maria Callas!

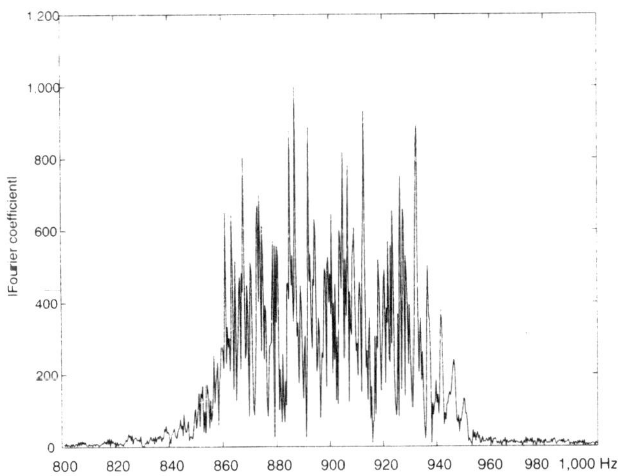

Figure 3.4. *Frequency analysis of A5 'interpreted' by Maria Callas, clearly showing the gap between the simple models we are describing and reality and its complexity...*

3.4. A brief history of A4

We owe the invention of the tuning fork to the luthier John Shore around 1711, more accurate than the 'tuning flutes' used before. The tuning fork's A4 has often changed over the centuries, but also from one place to the next, which was not without causing problems for traveling musicians! According to some, the tuning fork used by a city depended greatly on its financial state: if funds were low, the 'organ' budget was cut back by slightly shortening the pipes, and the tuning fork went up. Other instruments then had to play along... In 1704, the Paris opera's tuning fork was set to 405.3 Hz (below the current A♭), but gradually went up to 449 Hz in 1858. At the same time, it was at 434 Hz in London and 455.5 Hz in Brussels. Hence a decree in 1859 that set it in France to 435 Hz, which was confirmed by the Vienna congress of 1885. This did not stop it from continuing its senseless progression, and it can be tracked down again in London in 1953, where an international conference brings it to 440 Hz. In the meantime, the *Académie française* had solemnly but unsuccessfully attempted to bring it back to its former value of 435 Hz. Today, it wanders around under close watch between 440 and 444 Hz depending on the orchestras! But baroque music enthusiasts prefer to play it safe with the A 415, which is wiser with ancient string instruments...

3.5. Giving names to notes

In the early 11th century, Guy d'Arezzo sought to define a codification system for intervals. We owe him the names attributed to notes in Latin countries: Ut or *do* (C), *ré* (D), *mi* (E), *fa* (F), *sol* (G), *la* (A). The *si* (B) or *si*♭ will only later really appear. He is also responsible for the staff in musical notation. As a mnemonic device, he suggested a hymn to Saint John the Baptist attributed to Paul Diacre (around 770), in which the six first notes of the diatonic scale are found at the beginning of each verse:

UT queant laxis
REsonare fibris
MIra gestorum
FAmuli tuorum
SOLve polluti
LAbii reatum
Sancte Ioannes.

Today, the effectiveness of this mnemonic device is questionable! Letters were once used for notes, a practice that has lived on in Anglo-Saxon and German countries (see Table 3.1).

Latin notation	*do*	*ré*	*mi*	*fa*	*sol*	*la*	*si*♭	*si*	*do*
Anglo-Saxon notation	C	D	E	F	G	A	B♭	B	C
German notation	C	D	E	F	G	A	B	H	C

Table 3.1. *Names of the notes in different countries*

3.6. Other examples of scales

The simplification of the Western scale was probably imposed by polyphony. In other cultures where musical composition is more oriented towards melodic and rhythmic depth, scales are much more complex than ours, and are comprised of vast numbers of intervals with varying widths, quarter tones, three-quarter tones... with subtle differences that are difficult to perceive for our Western ears.

Figure 3.5. *Melodic formulas on the maquām* Rast. *The B$\flat$ and the E$\flat$ have to be played one quarter tone below their usual frequency*

The concept of scales is in fact too restrictive for Oriental music, and no term can render the exact meaning of the *maquām* of Arabic music, which simultaneously describes the intervals that are used, the movement of the melody (or the 'life of its own' [JAR 71]), the starting points, pauses, and final stops, all of this usually arranged inside a tetrachord (a sequence of four consecutive notes). The table 3.2 below indicates the intervals used in the *Rast*, *Hijāz* and *Saba* maquāms, taken from [JAR 71] where you will find associated melodies, including the one reproduced above.

C	D	E$\flat^+$	F	G	A	B$\flat$	C
D	E$\flat$	F$\sharp$	G	A	B$\flat^+$	C	D
D	E$\flat^+$	F	G$\flat$	A	B$\flat$	C	D (or E$\flat$)

Table 3.2. *From top to bottom, interval alterations in the Rast, Hijāz and Saba maquāms. The $\flat^+$ notation indicates a note lowered by a quarter tone*

3.7. Study problems

3.7.1. *Frequencies of a few scales (***)*

Fill in Table 3.3. The lowest F is F3. For all cases, take a C4 at 261.6 Hz as the reference frequency, and assume that the octaves are tuned without any beats. The 'well-tempered' scale is the one described in the chapter to play Bach. The one with inharmonicity will be calculated using the formula from Study problem 2.6.6, and by considering that B has the same value for every note: $B = 0.4/1\,200$, which corresponds to the case of a small upright piano.

3.7.2. *Beats of the fifths and the major thirds (*)*

Use the results of Study problem 3.7.1 to fill in Tables 3.4 and 3.5.

Scale:	F	F♯	G	G♯	A	B♭	B	C	C♯	D
Pythagorean										
Zarlino										
well-tempered										
equal temperament										
with inharmonicity										

Scale:	D♯	E	F	F♯	G	G♯	A	B♭	B	C
Pythagorean										
Zarlino										
well-tempered										
equal temperament										
with inharmonicity										

Table 3.3. *Frequencies for different scales*

Scale:	F-C	F♯-C♯	G-D	G♯-D♯	A-E	B♭-F
Pythagorean						
Zarlino						
well-tempered						
equal temperament						
with inharmonicity						

Scale:	B-F♯	C-G	C♯-G♯	D-A	D♯-A♯	E-B
Pythagorean						
Zarlino						
well-tempered						
equal temperament						
with inharmonicity						

Table 3.4. *Beats of the fifths for different scales*

Scale:	F-A	F♯-A♯	G-B	G♯-C	A-C♯	B♭-D
Pythagorean						
Zarlino						
well-tempered						
equal temperament						
with inharmonicity						

Scale:	B-D♯	C-E	C♯-F	D-F♯	D♯-G	E-G♯
Pythagorean						
Zarlino						
well-tempered						
equal temperament						
with inharmonicity						

Table 3.5. *Beats of the major thirds for different scales*

3.8. Practical computer applications

3.8.1. *Building a few scales*

Open the `play.m` file seen in Chapter 2 and save it under a new name, for example `scales.m`. In this new file, modify the vector `fr` (written as it is, it contains the frequencies of a slightly stretched tempered chromatic scale) so as to have it contain the frequencies of either the Zarlino or the Pythagorean scale, and listen to the differences between these scales and the tempered scale.

3.8.2. *Listening to beats*

We are going to listen to the beats produced between two notes. Open a new file and program a chord with two notes. This can be done either with the `instrument` function you created in the applications from Chapter 2 by writing in your file (after the necessary initializations):

```
s = instrument(f1,T,Fe) + instrument(f2,T,Fe);
```

or by using the `synthad` function (see applications from Chapter 1) by writing

```
s = synthad(a,fr1,p,T,Fe) + synthad(a,fr2,p,T,Fe);
```

In the first case, `f1` and `f2` are the fundamentals of the chord's two notes for which we wish to hear the beats. In the second case, the vectors `fr1` and `fr2` contain the list of harmonics, of the form `fr1 = (1:n)*f1` where `n` is the number of harmonics. You can try different values for the amplitudes contained in the vector `a`.

Try the following cases:

– `f1` and `f2` very close to each other;

– `f1` and `f2` slightly less than one fifth apart. Can you hear the beats? Can you still hear them if you remove the second and third harmonics?

– `f1` and `f2` set apart by one major third from the Pythagorean scale, the Zarlino scale, the tempered scale.

Conduct these tests for different values of pitch. Theoretically, what are the harmonics involved in the beats? What can you hear when you remove these harmonics? (This explains why flutists have such a hard time tuning their instruments when using thirds with very 'soft' flutes.) Add the fifth to it to compare the basic major chords.

Chapter 4

Psychoacoustics

Psychoacoustics is the study of the perception of sound. The sound processing performed by the ear and the brain is extremely complex, and its study [ZWI 81] is difficult because it involves subjectivity, as shown by the classification 'hearing, listening, understanding, comprehension' suggested by P. Schaeffer [SCH 66]. In this chapter, we will only be focusing on a few aspects of psychoacoustics: intensity and pitch, which are of direct interest to the musician, and the *masking* effects, of great use when designing sound compression techniques, such as the famous MP3[1] format described in Chapter 5.

4.1. Sound intensity and loudness

The sound intensity L_I that we defined in Chapter 1, usually expressed in decibels, is a physical measurement of acoustic pressure. However, this measurement does not coincide with our *sensation of sound intensity*, referred to as *loudness*, the study of which was developed by H. Fletcher in the 1930's [FLE 29]. First of all, we only hear sounds for a range of frequencies between 20 Hz and 20 kHz. But even inside this interval, for a given decibel level, loudness varies depending on the frequency. In particular, hearing shows a sensitivity maximum between 3,000 and 4,000 Hz (see Figure 4.1), allowing for example the piccolo to effortlessly stand out in a *tutti* orchestra. If you go back to Figure 1.17, you will also notice that the soprano's harmonics are more intense in that frequency range, so that her voice clearly stands out in the orchestral mass. There are two units of loudness to render the subjectivity of our hearing: the phon and the sone.

1. Short for MPEG Layer 3, which stands for moving picture expert group part 3 (audio).

4.1.1. *The phon*

By definition, the intensity in *phons* of a pure 1,000 Hz sound, hence of the form $s(t) = a\cos(2\,000\,\pi t + \theta)$, is equal to its measurement in decibels:

$$x \text{ dB} = x \text{ phons at } 1{,}000 \text{ Hz}. \tag{4.1}$$

Then, for a pure sound with any frequency f, its intensity in phons is by definition *the intensity in phons of the pure* 1,000 *Hz sound that would produce the same loudness.* This new measurement of intensity is denoted by L_N. Obviously, the above definition seems to depend on the listener, it is therefore necessary to conduct experiments on a large number of subjects, then to average the results, which led to establishing an international norm that precisely sets the relation between the levels L_I in dB and L_N in phons.

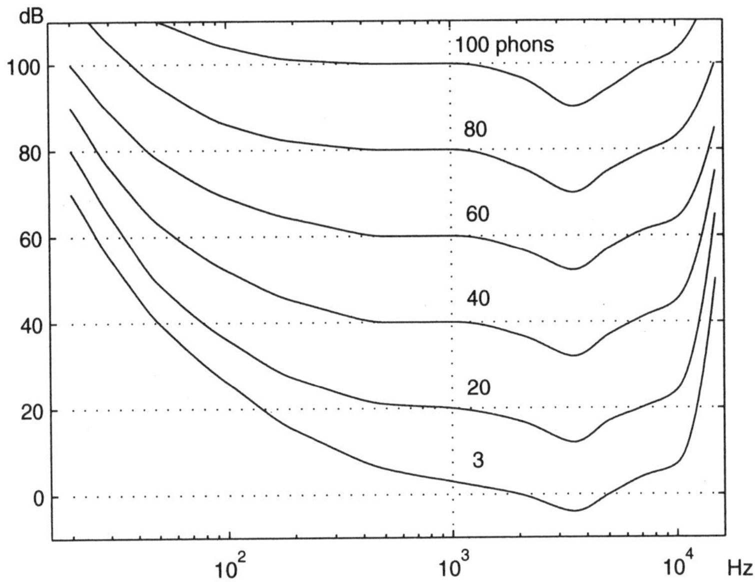

Figure 4.1. *Fletcher's equal-loudness contours. The ear's sensitivity maximum is located between* 3,000 *and* 4,000 *Hz*

Figure 4.1 shows a few *equal-loudness* contours, that is, lines along which the loudness of a pure sound is *constant*. Based on its definition, an equal-loudness contour

$$L_N = c \text{ phons}$$

includes the point with coordinates (1,000 Hz, c dB). The equal-loudness contour $L_N = 60$ phons, for example, tells us that the pure sound with frequency 100 Hz and intensity $L_I = 70$ dB, or the one with frequency 50 Hz and intensity $L_I = 80$ dB, produce the same sensation of sound intensity as a pure sound with frequency 1,000 Hz and with intensity $L_I = 60$ dB. The lower contour $L_N = 3$ phons marks the hearing threshold: any pure sound located below that line cannot be heard. Around 2,000 Hz and 5,000 Hz, the hearing threshold is 0 dB.

These contours were obtained for pure sounds. Studies and comparisons that led to similar results were conducted for other types of sounds (see [ZWI 81]): variable-band noise, periodic sounds, etc.

These studies showed that very low-pitched sounds (or also very high-pitched sounds) require more energy to be perceived with the same level of loudness. This explains, for example, why the baroque *basso continuo* is comprised of at least one bass such as the cello or the double bass, and one polyphonic instrument such as the harpsichord, both *simultaneously* playing the low-pitched part of the music.

4.1.2. *The sone*

The phon is related to the decibel by relation (4.1) and by the Fletcher contours. It fits the measurement of sound perception well in the sense that two pure sounds with different frequencies but the same measurement in phons are perceived to be at the same sound level. However, on its own, it provides no information as to the decibel level (or phon level) needed for a sound to be perceived as twice as loud as another sound. Fechner's law, which was stated in section 1.1.4, does tell us that the perception of intensity follows a logarithmic law, but in this case, this point of view deserves to be toned down [ZWI 81]. Many experiments conducted on pure 1,000 Hz sounds have shown that, on average, subjects find loudness to be doubled when the sound intensity is increased by 10 dB[2], which corresponds to multiplying the acoustic pressure by $\sqrt{10} \simeq 3.16$ (see section 1.1.4).

It is precisely this relation that the second unit of loudness conveys: the *sone*. An international agreement has set as the reference point the pure 1,000 Hz sound with intensity 40 dB and has attributed it with the loudness of 1 sone. The sound level in sones is denoted by N. Given the experiments mentioned above, the loudness level of a pure 1,000 Hz sound is therefore

$$N = 2^{(L_N - 40)/10}$$

2. To be more precise, this is valid when the weaker sound is above 40 dB. Below 40 dB, the gap that leads to a twofold sensation steadily grows from 3 to 10 dB.

To then find the level in sones of a pure sound with any frequency, we simply have to refer to Fechner's equal-loudness contours. Using the same example as above, the pure sound with frequency 100 Hz and intensity $L_I = 70$ dB or $L_N = 60$ phons therefore has a loudness level $N = 4$ sones.

Assuming that the above applies to other sounds, we can make the following comments: in a concert hall, the background noise is close to 40 dB, or roughly 1 sone, and a *tutti* orchestra can reach 110 dB. The eight degrees of intensity *ppp, pp, p, mp, mf, f, ff, fff* are more or less equivalent to 40, 50, 60, ..., 110 dB, or 1, 2, 4, ..., 128 sones. We saw in sections 1.1.4 and 1.6.3 that multiplying the number of instruments by ten was equivalent to a 10 dB increase in sound intensity. Hence we have to multiply the number of instruments by ten (or their amplitudes by $\sqrt{10}$[3]) to multiply the loudness by two. Rather than a logarithmic law, this is a fractional exponent law:

$$N \simeq a n^{\alpha},$$

where n is the number of instruments, $\alpha = \log 2 \simeq 0.301$ and a is a constant that depends on the instrument.

4.2. The ear

To understand how we can tell different frequencies apart, with a resolution of about five *cents* (see section 2.5.1.1) between 300 and 3,000 Hz, a little anatomic field trip into the ear is in order (Figures 4.2 and 4.3). Its key vibrating element is the *cochlea*, a thin tube, roughly 32 mm long, filled with lymph, attached to the side of a thin strip of bone and partly bounded by two membranes, the basilar membrane and Reissner's membrane. This tube is lined on its length with four rows each comprised of 3,500 sensors: these are the *hair cells*, which send electrical impulses to the brain through the acoustic nerve. Each cell can reach an estimated maximum rate of 1,000 discharges per second, not enough to account for the discrimination of sounds with frequencies above 500 Hz (see section 5.1). It is the combined action of the many hair cells that enable such a discrimination.

By assigning the value 1 to each electrical impulse of a hair cell (0 indicating the absence of an impulse), and by assuming that the $4 \times 3{,}500 = 14{,}000$ cells can simultaneously provide 1,000 impulses per second, we end up for the bitrate of one ear (hence in 'mono') with the impressive value of 14 Mbit/s! In comparison, the mono track of an audio CD sampled at 44 kHz using 16 bits (see Chapter 5) has a bitrate of 0.7 Mbit/s, or twenty times less than the ear's estimated maximum rate. This

3. It may seem strange not to have the same multiples for the amplitude and the number of instruments. This is due to the fact that the phases of the instruments are usually randomly shifted with respect to one another, partly cancelling out their contributions (see Study problem 1.6.3).

is an indication of the ear's good performance, but also of the fact that increasing the audio quality of CD's by a factor higher than 20 should not lead to a noticeable improvement.

The cochlea's section shrinks as it goes from the elliptical window to its extremity, the helicotrema, whereas the basilar membrane grows wider instead. The complete mechanism is difficult to analyze from a mechanical point of view, but by direct observation, V. Békésy noticed around 1960 that the amplitudes of the cochlea's vibrations produced by the high frequencies reach their maximum in the part close to the elliptical window, whereas for low frequencies, the amplitudes reach their maximum at the extremity [ZWI 81] (see Figure 4.3). Back in the 19th century, the physiologist and physician H. Helmholtz foresaw this localization of frequencies [HEL 68, FIC 96]. He believed that each hair cell, like a piano string, was associated with a specific frequency.

So what you should remember is that for each sound, a 'sound signature' is produced on the cochlea, and sent to the brain by the hair cells. All of the ear's other elements are there to work as a medium between the outside and the cochlea and between the cochlea and the brain:

– the outer ear picks up sound waves through the pinna and carries them through the auditory canal down to the tympanic membrane (the eardrum);

– the middle ear is comprised of a mechanism meant to decrease the amplitude of the vibrations while increasing their efficiency (leverage effect), ensuring the air-liquid transmission of these pressure variations: this is the chain of ossicles, which is stimulated by the eardrum, and transmits the vibrations to the elliptical window;

– the inner ear, shaped like a snail shell coiled about two and a half times, is a tube containing the cochlea floating in lymph, which is set into vibration by the elliptical window;

– the acoustic nerve transmits the information from the hair cells to the brain.

4.3. Frequency and pitch

The sensation of pitch is of course related to the frequency. To a first approximation, Fechner's law still applies: pitch varies with the logarithm of the frequency. So the gap we perceive between a 100 Hz sound and a 200 Hz sound is the same as between a 200 Hz sound and a 400 Hz sound: this is the octave interval, equivalent to a twofold increase in frequency. The interval measured in octaves between two pure sounds is therefore equal to the difference of the base 2 logarithms of their frequencies. But this no longer applies for high frequencies, and a pure 6,000 Hz sound seems much lower than the octave of a pure 3,000 Hz sound.

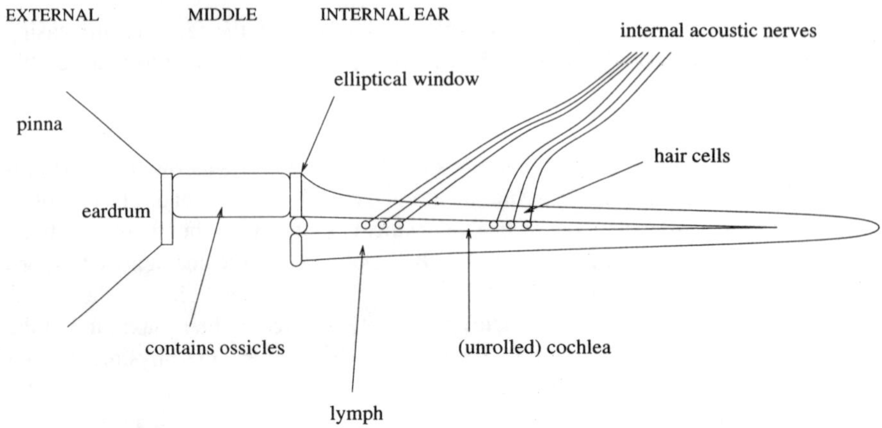

Figure 4.2. *Diagram of how the ear works. The essential organ, the cochlea, bathes in an aqueous solution (a reminiscence of our past as fish?). The ossicles act as a lever to ensure the air-liquid medium change*

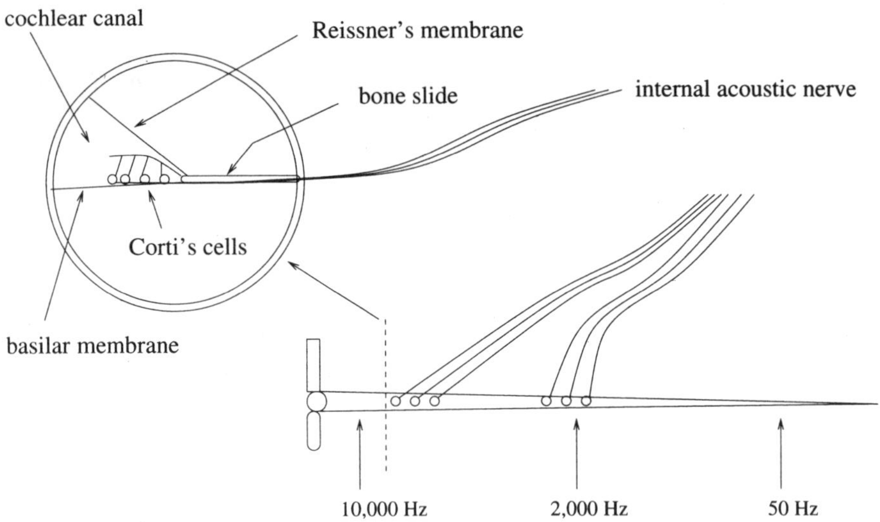

Figure 4.3. *Cross section of the cochlea and localization of the frequencies*

4.3.1. *The mel scale*

The *mel* scale (or Stevens scale) is meant to account for this distortion. By definition, a pure sound of 125 Hz (or 131, or 1,000,... depending on the source) is attributed 125 mels (or again, another value, depending on the source), then, through experiments conducted on a large number of subjects, the mel scale is calibrated so that a pure sound with $2x$ mels gives the sensation that it is exactly one octave above a pure sound with x mels. For example, the sequence of octaves 500, 1,000, 2,000 mels is more or less equivalent to the frequencies 500, 1,010, 2,050 Hz.

However, the relevance of this scale is questionable for at least two reasons: the high variability among individuals, and the fact that it only applies to pure sounds, which are virtually non-existent when it comes to acoustic instruments. Nonetheless, this is probably the reason behind a common practice among violinists and piano tuners who tend to widen the upper octaves, and it also provides some theoretical justification for Cordier's temperament with perfect fifth [COR 82] (see section 3.3). Maybe it could also explain Maria Callas' A5 (Figure 3.4)? And for those who enjoy music dictation, what notes can you hear in the sound illustration 'the mountain whistler' on the AM website, the spectrogram of which is shown in Figure 1.20?

Furthermore, it is worth noting that pitch also varies with intensity: low-pitched sounds seem lower when their intensity increases, whereas high-pitched sounds seem higher, with an apparent pitch variation of up to an entire tone when the sound intensity goes from 40 dB to 100 dB, for frequencies around 150 Hz or 5,000 Hz. For 2,000 Hz sounds, the pitch variation is insignificant.

4.3.2. *Composed sounds*

4.3.2.1. *Pitch of sounds composed of harmonics*

For sounds comprising several harmonics, the sensation of pitch is usually provided by the frequency of the fundamental. Its presence is actually not essential: to be convinced, simply listen to some music on a small radio set that cannot deliver any frequency below for example 150 Hz. Even so, a listener can perfectly recognize the notes that are played, even if the fundamentals of some notes are missing. To identify a note with fundamental f, the presence of a few harmonics multiples of f is often sufficient [LEI 80]. Removing the low-pitched harmonics, however, can lead to the sensation of a slightly higher-pitched sound, whereas removing the high-pitched harmonics can lead to the sensation of a slightly lower-pitched sound. There lies maybe the explanation behind the endless conflict between musicians who accuse each other of playing out of tune, particularly when some are playing backstage, which can filter out certain harmonics.

4.3.2.2. *Pitch of sounds composed of partials*

For sounds composed of partials, the concept of pitch can become rather uncertain, as is the case for drums. The same sound can be perceived with a different pitch depending on the individual, because not everybody will necessarily hear the same partials, or interpret their frequency ratios the same way. This ambiguity also occurs sometimes with bells. Usually, the second partial, called the principal, gives the bell's pitch, and not the first, located one octave below and referred to as the *hum*. Because of still the same ambiguity, you may not necessarily all agree on the highest note of our mountain whistler (see AM website): is it located between D8 and E♭8 as the first partial (about 4,910 Hz) would suggest? Or between A7 and B7 if we consider the second partial (about 7,320 Hz)? And finally, don't we instead hear an intermediate note: a C8?

4.3.3. *An acoustic illusion*

We end this section with a beautiful acoustic illusion on pitch: the perpetually ascending sound, synthesized on a computer by Shepard [RIS 99], and not without some resemblance with reprise in organs (see section 2.5.1.3).

The idea is to arbitrarily choose a scale, for example the whole tone scale (with a thought for Claude Debussy), and to progressively add low harmonics to each note, while at the same time removing the upper harmonics.

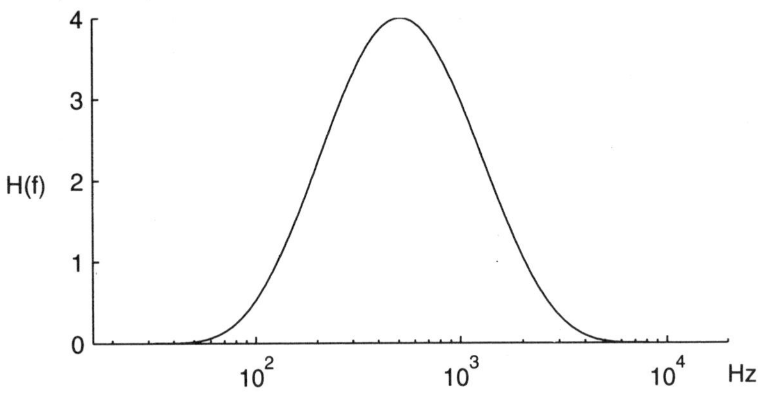

Figure 4.4. *Harmonics truncation function*

This is done by defining a function H equal to 0 outside the interval $[32\text{ Hz}, 8{,}192\text{ Hz}]$ (the base 2 logarithms of the boundaries are 5 and 13) and with its maximum around an average frequency, for example the following function (see Figure 4.4):

$$H(f) = \begin{cases} (1 + \cos[\pi(2\log_2(f) - 18)/8])^2 & \text{if } 32 \leq f \leq 8\,192, \\ 0 \text{ otherwise.} \end{cases}$$

Consider the frequencies $f_k = 220 \times 2^{k/6}$, $k = 0, 1, 2, ...$, that make up the ascending whole tone scale. The *perpetually ascending scale* is then comprised of the following periodic sounds (note that, by definition of H, the sums are actually finite):

$$N_k(t) = \sum_{n=-\infty}^{+\infty} H(2^n f_k) \sin(2\pi 2^n f_k t).$$

The gap between any two consecutive harmonics of the note N_k is an octave. Also, the harmonics of N_{k+1} are all located exactly one tone above their equivalent harmonics of N_k, since they have the frequency ratio $f_{k+1}/f_k = 2^{1/6}$. Therefore, N_{k+1} will be perceived as being one tone higher than N_k. But, by definition,

$$N_6 = N_0.$$

And likewise $N_{6+m} = N_m$ for any m! We therefore have the following paradoxical situation: even though each note is higher than the previous one, we end up with the same note six notes later, exactly like in the famous drawing by Escher showing a closed-loop staircase! The spectrogram of the first eight notes is shown in Figure 4.5. In 1968, Risset built a 'continuous' perpetually ascending sound, based on the same idea, in other words a *glissando* [RIS 99].

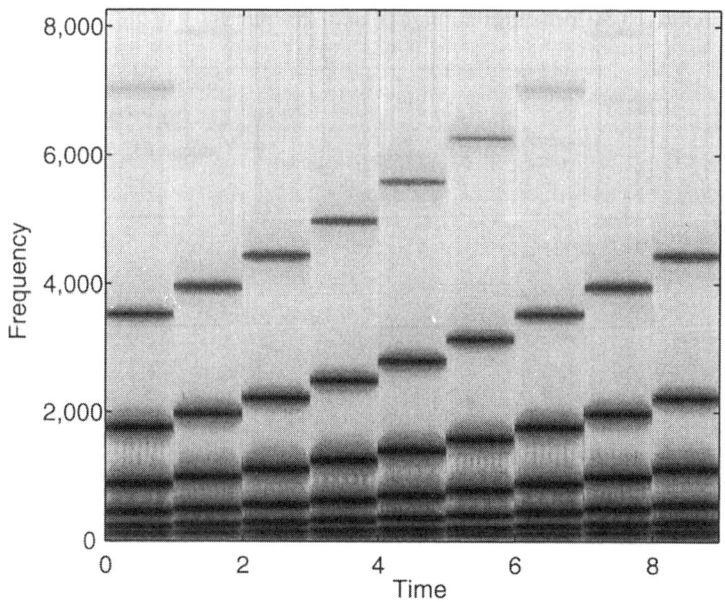

Figure 4.5. *Endlessly starting over...*

4.4. Frequency masking

When out of two sounds produced simultaneously, or almost simultaneously, only one of them is perceived by the listener, the other one is said to be *masked* by the first. Masking effects are extensively studied in [ZWI 81]: time masking, where an intense sound masks a fainter sound that follows it, or even precedes it, masking of a pure sound by white noise, masking of different variable bandwidth noises, etc. The advantage of having a good model for masking effects is that they can be used to develop sound compression algorithms, based on the simple following principle: *there is no point in keeping what the listener will not perceive!* Therefore, we can neglect all of the masked sounds, and thus gain on the volume of the data representing the sound. Particularly, this makes it possible to increase the bitrate of cable or satellite transmissions. Compression techniques are examined in Chapter 5.

Here, we will be studying the simple but interesting case of a pure sound masked by another pure sound with different frequency and intensity, both sounds being produced simultaneously. The typical procedure for experimentally measuring the masking effect is as follows: the intensity of the masking sound is maintained constant, and starting with a zero intensity for the masked sound, the subject is asked to gradually turn the dial that controls its intensity until it becomes audible. By conducting a large number of these experiments, the result is an average that can be used as a model. The experiment is repeated with different frequencies for the masking sound and the masked sound to cover the entire range of audible sounds.

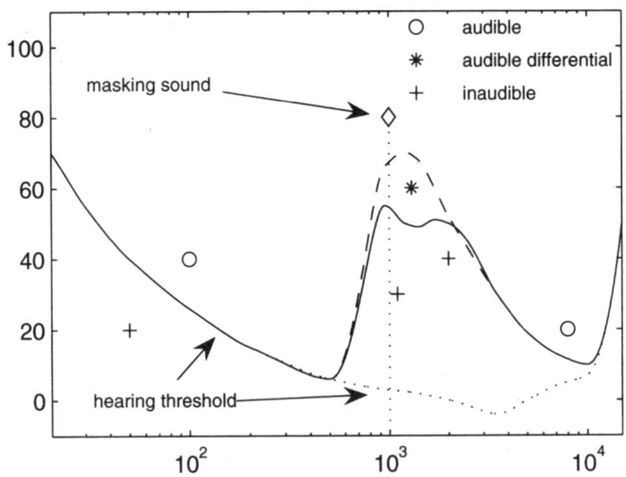

Figure 4.6. *Frequencies masked by a pure* 1,000 *Hz sound at* 80 *dB*

Figure 4.6 shows the *masking threshold*, or *mask*, produced by a masking sound with frequency 1,000 Hz and intensity 80 dB. This is the full line. Any sound below this line is masked, hence inaudible. For certain sounds, those above this line but below the dashed line, the masked sound cannot be heard, but on the other hand, the listener can perceive the *differential sound* produced by the beat (see Figure 2.3) between the two sounds. For example, if the masked sound has a frequency of 1,200 Hz and an intensity of 60 dB, the differential sound that is heard has a frequency of 1 200 Hz − 1 000 Hz = 200 Hz. The shape of the mask varies both with the pitch and the intensity of the masking sound. Figure 4.7 shows the masking thresholds for different intensities L_I of the masking sound, still with a frequency of 1,000 Hz.

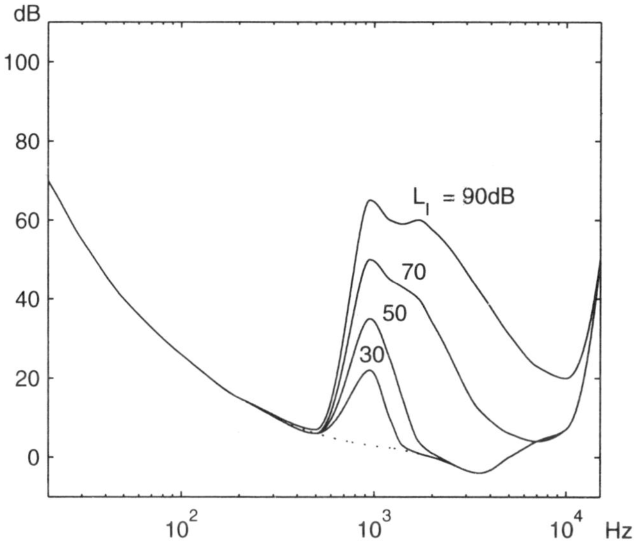

Figure 4.7. *Masking effect for a* 1,000 *Hz masking sound, with variable intensity* L_I

The masking threshold lines show a rather clear dissymmetry: an intense low-pitched sound easily masks a faint high-pitched sound, whereas an intense high-pitched sound will have more difficulty masking a low-pitched sound. One possible explanation [SON 79] lies in the frequency localization (see Figure 4.3): in order to stimulate the extremity of the cochlea, low frequencies have to go through the area that receives high frequencies, and therefore can have an effect on the perception of these high frequencies. On the other hand, high frequencies, located at the beginning of the cochlea, near the elliptical window, do not affect the extremity of the cochlea where low frequencies are picked up, and it is therefore not surprising that they have no effect on these low-frequencies. By remembering that women sing roughly one octave above men, Zwicker sees in this dissymmetry the explanation to the fact that there are less men than women in choirs. No, it isn't a socio-cultural phenomenon!

4.5. Study problems

4.5.1. *Equal-loudness levels (**)*

Fletcher's equal-loudness contours were obtained by averaging results obtained for large numbers of people. Design an experimentation protocol and draw the equal-loudness contours for your own hearing.

4.5.2. *Frequency masking (**)*

Design an experimentation protocol and draw the equal-loudness contours for your own hearing, for the following masking frequencies: 200 Hz, 1,000 Hz and 3,000 Hz.

4.5.3. *Perpetually ascending sound (**)*

Based on the example of Shepard's ascending scale, build a sound that seems to *glissando* up indefinitely, while periodically going through the same values.

4.6. Practical computer applications

4.6.1. *Frequency masking*

Write a program that overlaps a pure 1,000 Hz sound (the *masking* sound) and a pure sound with frequency 1,010 Hz (the *masked* sound), the latter successively assuming the relative sound levels of −35 dB, −30 dB, −25 dB, −20 dB, −15 dB with respect to the former. At what level do you begin to perceive the second sound?

Same questions for a masked sound with frequency:
1,100 Hz and with relative sound levels of −35 dB, −30 dB, −25 dB, −20 dB, −15 dB,
1,500 Hz and with relative sound levels of −35 dB, −30 dB, −25 dB, −20 dB, −15 dB,
2,500 Hz and with relative sound levels of −55 dB, −50 dB, −45 dB, −40 dB, −35 dB.

4.6.2. *Perpetually ascending scale*

Use the formulas from this chapter to program the perpetual whole tone scale. To do this, create a function with the same format as the `instrument` function already created, named for example `noteps`, and which produces a sound s, the harmonics distribution of which follows the suggested model. To create a sequence of $N+1$ notes set one tone apart, you can write a loop of the form

```
for k = 0:N
   f = f0*tone^k;
   s = [s, noteps(f,T,Fe)];
end
```

where `tone` is the ratio between two consecutive tones and `f0` is the 'frequency' of the first note.

Chapter 5

Digital Sound

Generally speaking, an *analog* signal is a signal produced by a mechanical or electronic device. For such a signal, the variable is time, which elapses *continuously*. Just a few decades ago, any sound production chain was completely analog. For example, the sound produced by the musicians, the electrical signal delivered by the microphones, the signal transmitted by radio waves or engraved onto a phonograph record, the signal received and amplified by your stereo system and finally the sound produced by the speakers, all these are analog signals.

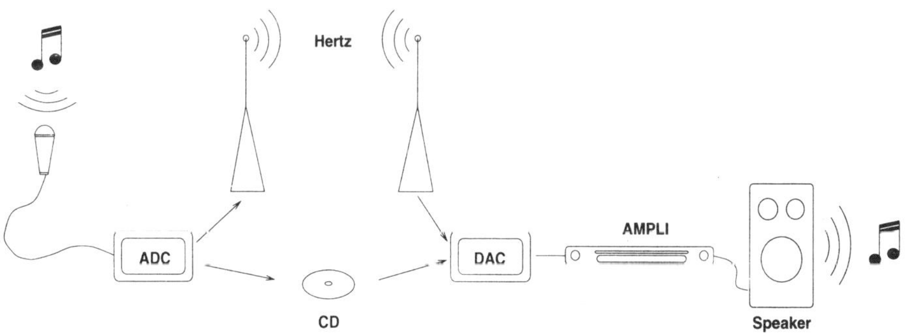

Figure 5.1. *Digital audio chain: the inevitable use of the analog-to-digital convertor (ADC), and after that of the digital-to-analog convertor (DAC)*

With the tremendous advances in computer capabilities, a new link has appeared in the chain: digital sound. Once the sound is captured by the microphone, it is transformed into a sequence of *binary numbers* (made of 0's and 1's), which are transmitted, stored or engraved in that form. The device that operates the conversion is called an *analog-to-digital converter* (ADC). It actually performs two distinct tasks on the analog signal $s(t)$:

– the *sampling*, which consists of measuring the values $s_n = s(n\tau)$ of the analog signal at regularly spaced intervals of time 0, τ, 2τ, 3τ, ... where τ is called the *sampling period*. The standard *sampling frequency* $F_e = 1/\tau$ for audio CD's is 44.1 kHz, or 44,100 values per second;

– the *quantization*, which consists of approximating and replacing these real numbers s_n, which can have an infinite number of decimals that would be impossible to store, with numbers r_n taken from a *finite* set of $L = 2^b$ possible values. Each of these numbers r_n is then encoded using b bits[1] in order to be stored or transmitted. For audio quality, generally 16-bit encoding is used, which is equivalent to 2 bytes.

The resulting signal is a *digital* signal. It has no physical reality beyond the numbers that comprise it, somewhere in the computing universe, waiting to be processed by a sound card. It is made up of the sequence of numbers r_n, where the variable now is the integer n: it is a *discrete* variable. To reconstruct the sound, the digital signal is converted back to an electrical analog signal by a *digital-to-analog converter* (DAC), and proceeds through the rest of the chain's usual components.

The question that spontaneously comes to mind is this: shouldn't there be a loss of information caused by the conversion to a digital signal? We will see that under certain hypotheses, there is no loss. Unfortunately, these hypotheses are never verified and never will be: we are faced with a fundamental theoretical obstacle; but we can be 'close'! Everything lies in that close, an unavoidable imperfection (insert favorite saying here), but endlessly reduced by technological advances.

When trying to reduce the error caused by the digital process, the price to pay lies in the large quantity of data obtained for representing the sound: currently, an hour long stereo recording sampled at 44.1 kHz with 2 byte encoding takes up $3{,}600 \times 2 \times 44{,}100 \times 2 = 635$ MB (megabytes) on an audio CD. We will see that it is possible to reduce the size of the data using the psychoacoustic properties of hearing, which can be very useful, particularly for Internet transmissions. Of course, this compression usually comes with a loss of information, which increases with the extent of the compression. But the algorithms involved are designed so that the lost data, as much as possible, is precisely the data that would not have been heard.

We will finish this chapter with a few concepts of digital filtering, and establish the connection with analog filtering. These concepts will be of use to us in Chapter 6, particularly in regard to sound effects.

1. A bit is a binary digit, equal to 0 or 1. A byte is composed of 8 bits.

5.1. Sampling

Consider a sound $s(t)$, where the function s is continuous and bounded on $\mathbb{R}$. After a sampling period $\tau > 0$ has been chosen, the *sampled sound* consists of a sequence of values or *samples* taken at the instants $t_n = n\tau$:

$$s_n = s(t_n), \quad n = \ldots, -1, 0, 1, 2, \ldots$$

Figure 5.2 shows a sound with a duration of 0.01 s, sampled at 2,000 Hz. Aside from the quantization (see section 5.1.2), the values s_n are those that are stored on an audio CD.

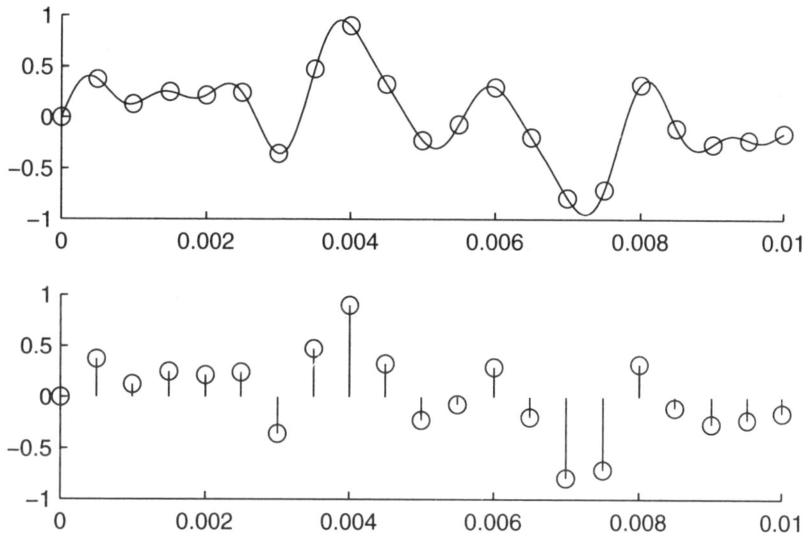

Figure 5.2. *Initial sound $s(t)$ (top) and sampled sound $s_e(t)$ (bottom)*

A more elaborate description of the sampled sound, abundantly used in signal theory, consists of representing this sound with an infinite number of *Dirac impulses* $\delta_{n\tau}(t)$, each one located at the instant $t_n = n\tau$ and with a *mass* equal to the quantity τs_n, where the purpose of the factor τ is simply to bring these terms to scale (see equation (5.1)). We have already come across the Dirac impulse at point 0 (see relation (1.22)). Generally speaking, the Dirac impulse at point a and with mass $\mu \in \mathbb{C}$, denoted by $\mu\delta_a(t)$, is such that

$$\int_{-\infty}^{+\infty} \varphi(t)\mu\delta_a(t)\, dt = \mu\varphi(a)$$

for any continuous function φ. To get a better idea of what this Dirac impulse represents, you can consider the sequence of rectangular functions (see Figure 5.3, where $a = 2$ and $\mu = 1$) defined for $n \geq 1$ by

$$u_n(t) = \begin{cases} \mu n \text{ if } |t-a| < 1/(2n), \\ 0 \text{ otherwise.} \end{cases}$$

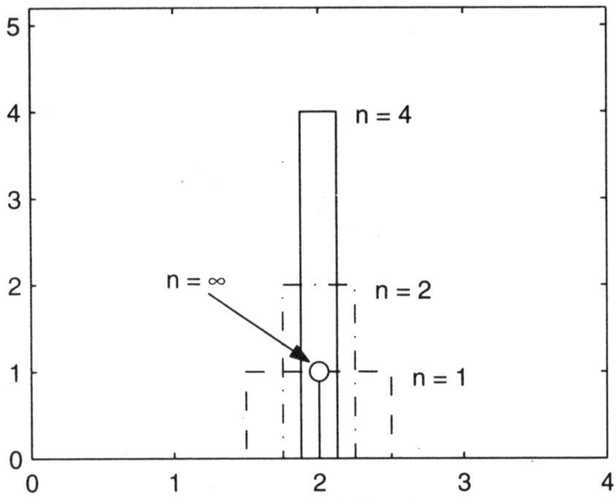

Figure 5.3. *Sequence of functions $u_n(t)$ the limits of which are the Dirac impulse at point $a = 2$ with mass $\mu = 1$, denoted by the vertical line and the small circle with x-coordinate a and y-coordinate μ*

These functions are focused more and more at point a, and have the same 'mass' μ since

$$\int_{-\infty}^{+\infty} u_n(t)\, dt = \mu.$$

What we get at the limit is not a function (it is called a *distribution*), but instead is precisely the Dirac impulse $\mu\delta_a(t)$:

$$\lim_{n\to\infty} \int_{-\infty}^{+\infty} u_n(t)\varphi(t)\, dt = \mu\varphi(a) = \int_{-\infty}^{+\infty} \varphi(t)\mu\delta_a(t)\, dt.$$

The sampled sound is then defined by

$$s_e(t) = \tau \sum_{n=-\infty}^{+\infty} s_n\, \delta_{n\tau}(t) \tag{5.1}$$

and its graphical representation (except for the factor τ) is given at the bottom of Figure 5.2, according to the rule that circles have the mass of the associated Dirac impulse as their y-coordinate.

5.1.1. *The Nyquist criterion and the Shannon theorem*

The fundamental question is to determine under what condition the original sound can be reconstructed from no more information than its samples. At first, the task seems impossible: Figure 5.2 clearly shows that the values of $s(t)$ between two samples can be modified without changing the values of these samples. Therefore, two distinct sounds with the same samples will lead to the same electrical signal at the DAC's output, since the DAC only takes the samples as its input and nothing else. At least one of the two reconstructed signals will be different from the original signal!

5.1.1.1. *Case of a sinusoidal signal*

To understand what limitations have to be imposed on the sound $s(t)$ in order to reconstruct it exactly through the ADC-DAC chain, we will first examine the case of a sinusoidal signal

$$s(t) = \alpha \cos(2\pi f t + \theta).$$

Considering the fact that we are dealing with a sinusoidal signal, but with any possible amplitude, frequency and phase, what is the minimum sampling frequency needed to reconstruct $s(t)$ from nothing more than its samples? We could for example suggest taking a sample every time $s(t)$ reaches a maximum, which would lead to one sample per period, but this would not be enough to tell the difference between an oscillating signal and a constant signal (see Figure 5.4).

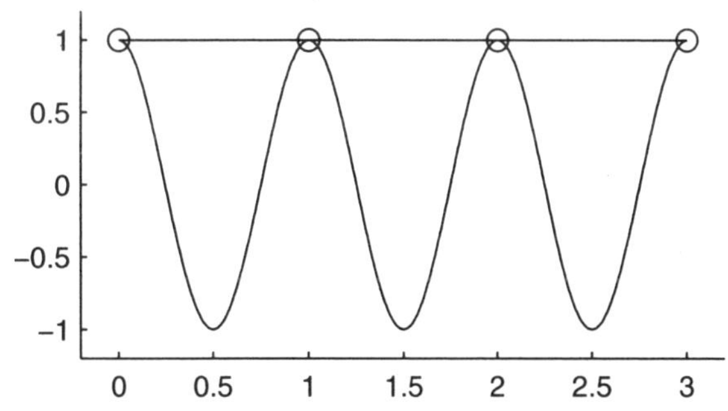

Figure 5.4. *Cosine function and constant signal (equal to 1) that produce the same samples*

Then we can suggest taking at least one sample every time $s(t)$ reaches a maximum or a minimum, which leads to *at least two samples per period*. And we still have to avoid ending up with the instants where $s(t) = 0$, which happens every half-period.

Knowing that the period of the signal $s(t)$ is equal to $1/f$, this leads us to considering the following hypothesis regarding the sampling period:

$$\tau < \frac{1}{2f}.$$

Because the sampling frequency is $F_e = 1/\tau$, this condition is equivalent to

$$F_e > 2f, \tag{5.2}$$

which is called the *Nyquist criterion* or the *Shannon condition.*

We will now provide a more rigorous justification for this condition. Switching over to complex numbers makes it much more convenient to make the following argument. Remember that this means negative frequencies also have to be considered, and the above criterion becomes

$$F_e > 2|f|. \tag{5.3}$$

So we ask ourselves this question: if two harmonic signals

$$s_1(t) = c_1 \exp(2i\pi f_1 t),$$

$$s_2(t) = c_2 \exp(2i\pi f_2 t),$$

produce the same samples, are they equal? First we infer from $s_1(0) = s_2(0)$ that $c_1 = c_2$, and we define $c = c_1$. If $c = 0$, then $s_1(t) = s_2(t) = 0$ for any t, and the two signals are equal. Now let us assume $c \neq 0$. The fact that the two signals are equal at $t = \tau$ leads to

$$c \exp(2i\pi f_1 \tau) = c \exp(2i\pi f_2 \tau),$$

hence $\exp(2i\pi(f_1 - f_2)\tau) = 1$ and $(f_1 - f_2)\tau$ is an integer, meaning that

$$f_1 - f_2 = \frac{k}{\tau} = kF_e, \quad k \in \mathbb{Z}. \tag{5.4}$$

But, if we assume that the two frequencies f_1 and f_2 meet the Nyquist criterion (5.3), we have

$$|kF_e| = |f_1 - f_2| \leq |f_1| + |f_2| < \frac{F_e}{2} + \frac{F_e}{2} = F_e$$

with k an integer, which imposes $k = 0$, and therefore $f_1 = f_2$. On the other hand, if the Nyquist criterion is not imposed, k can be chosen different from 0, and in that case $f_1 \neq f_2$. Also, by observing the following samples, we notice that

$$s_1(t_n) = c \exp(2i\pi f_1 n\tau) = c(\exp(2i\pi f_1 \tau))^n$$

$$= c(\exp(2i\pi f_2 \tau))^n = c \exp(2i\pi f_2 n\tau) = s_2(t_n),$$

meaning that we indeed have the same samples for two different signals.

Thus, the Nyquist criterion is necessary and sufficient for two harmonic signals that produce the same samples to be equal.

5.1.1.2. *General case*

What we just saw is a particular case of the fundamental result in communication theory: the Shannon theorem [SHA 49]. Before we state it, we need the following definition: let B be such that $B > 0$, and let $s(t)$ be a signal that has a Fourier transform $\widehat{s}(f)$. Remember that we have the following expression (see formula (1.14)):

$$s(t) = \int_{-\infty}^{+\infty} \widehat{s}(f) \exp(2i\pi ft)\, df,$$

meaning that $\widehat{s}(f)$ is the frequency density of f in the signal $s(t)$. A signal $s(t)$ is said to be a $[-B, B]$ *band-limited signal* if

$$\widehat{s}(f) = 0 \text{ for any } |f| > B,$$

or, in other words, if the signal contains no frequencies such that $|f| > B$. For example, although we have not defined the Fourier transform of $s(t) = \exp(2i\pi ft)$, the frequency of this signal is f, hence it is $[-|f|, |f|]$ band-limited.

Shannon theorem. *Let $s(t)$ be a function that has a $[-B, B]$ band-limited Fourier transform $\widehat{s}(f)$. This function is sampled at the frequency F_e. If F_e meets the Nyquist criterion*

$$F_e > 2B,$$

then $s(t)$ is the only $[-B, B]$ band-limited function with the values $s(n\tau)$, $n \in \mathbb{Z}$ as its samples, with $\tau = 1/F_e$. Additionally, if $h(t)$ is the ideal low-pass filter (1.26) with cut-off frequency $F_e/2$, then $s(t)$ can be reconstructed from the sampled signal $s_e(t)$ given by (5.1) by passing it through the filter:

$$s(t) = (h * s_e)(t).$$

Skeptics could argue that this seems absurd: what happens if the values of $s(t)$ are modified between the samples without changing the samples? The reconstruction will not work... What happens is this, something the theorem also states: it is *impossible to modify* the values $s(t)$ *between* the samples without modifying them, while still maintaining the $[-B, B]$ band-limited condition. Any modification will necessarily introduce frequencies higher than B, and the theorem no longer applies. So the reconstruction will in fact not work, but only because the hypotheses of the theorem are no longer satisfied.

5.1.1.3. *Consequences*

What are the implications of this result in a practical case? Our ears cannot perceive frequencies higher than 20 kHz. It is therefore useless during a sound recording to record sounds with frequencies higher than 20 kHz. This can be achieved (more or less, see next section) either by a low-pass filtering of the recorded signal, or simply because the microphone is not sensitive enough to pick up ultrasounds. With $B = 20$ kHz, the conditions are thus met for the theorem, which states that if we sample the signal at a frequency higher than $2B = 40$ kHz, then the original signal (or to be more precise, the signal that was stripped of its frequencies higher than 20 kHz) can be exactly reconstructed by filtering the sampled signal with the appropriate low-pass analog filter. This is why the frequency generally used for *audio quality* is chosen above 40 kHz.

5.1.1.4. *Theoretical impossibility*

This is all very interesting, but there's a hitch: aside from the zero function, *there is no* function equal to zero outside a finite interval with a Fourier transform that is also equal to zero outside a finite interval (see section 1.4.2). Oh really? So what? Well, in the Shannon theorem, we assumed that $\widehat{s}(f)$ was equal to zero outside $[-B, B]$. So $s(t)$ cannot be equal to zero outside a finite interval. There lies the whole problem: the theorem only applies to a signal with an infinite 'duration', which is never the case in music: any recorded sound has a beginning and an end! And therefore, it inevitably contains arbitrarily high frequencies... Which brings us to the following question.

5.1.1.5. *What happens if the Nyquist criterion is not met?*

Let us see that with an example. For a medium audio quality, our sampling frequency will be $F_e = 30$ kHz, with period $\tau = 1/F_e$, and we will sample the pure sound

$$s(t) = 2\cos(2\pi f t) = \exp(2i\pi f t) + \exp(-2i\pi f t),$$

with frequency $f = 27$ kHz, absolutely inaudible (ultrasound).

Based on the analysis we did in section 5.1.1 and particularly on (5.4), $\exp(2i\pi f t)$ produces the same samples as $\exp(2i\pi f_1 t)$ with $f_1 = f - F_e = -3{,}000$ Hz, and $\exp(-2i\pi f t)$ produces the same samples as $\exp(2i\pi f_2 t)$ with $f_2 = -f + F_e = -f_1$. By a simple addition (because sampling is linear), $s(t) = \exp(2i\pi f t) + \exp(-2i\pi f t)$ therefore has the same samples as

$$\begin{aligned} v(t) &= \exp(2i\pi f_1 t) + \exp(2i\pi f_2 t) \\ &= \exp(-2i\pi f_2 t) + \exp(2i\pi f_2 t) \\ &= 2\cos(2\pi f_2 t) \end{aligned}$$

with $f_2 = 3{,}000$ Hz. This means that we have $v(n\tau) = s(n\tau)$ for any $n \in \mathbb{Z}$, in other words

$$v_e(t) = s_e(t).$$

Other frequencies, of the form $\pm(f - kF_e)$, would also lead to the same samples (see Figure 5.5). Let us then assume for example that $B = 14$ kHz. Because $F_e > 2B$ and because the pure sound $v(t)$ is $[-B, B]$ band-limited, the Shannon theorem applies to v, and therefore, the signal reconstructed by filtering s_e with an ideal low-pass filter h with bandwidth $[-F_e/2,\ F_e/2]$ will be exactly the sound $v(t)$:

$$(h * s_e)(t) = (h * v_e)(t) = v(t).$$

We have just gone from a pure inaudible 27 kHz sound to a pure 3,000 Hz sound, where our hearing is the most sensitive!

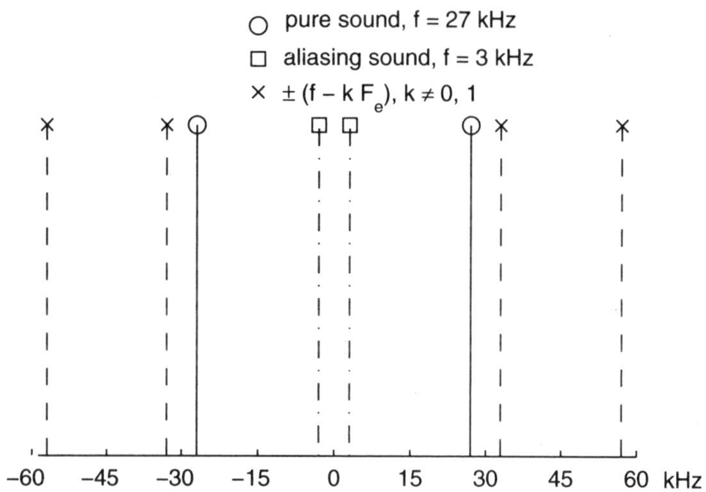

Figure 5.5. *For a given frequency f, and for example $F_e = 30$ kHz, other frequencies lead to the same samples. Watch out for those contained in the $[-F_e/2, F_e/2]$ band!*

This effect is called *aliasing*. 3 kHz is the *symmetric* point of 27 kHz with respect to half the sampling frequency $F_e/2 = 15$ kHz. It is as if Figure 5.5 had been folded along the vertical axis with x-coordinate $F_e/2$. The same folding occurs in fact along the axis with x-coordinate $-F_e/2$.

The argument we just made is valid in the general case. What happens in reality is that recorded sounds are not quite $[-20$ kHz, 20 kHz$]$ band-limited (because of the theoretical impossibility mentioned above), but have a low frequency density beyond 20 kHz.

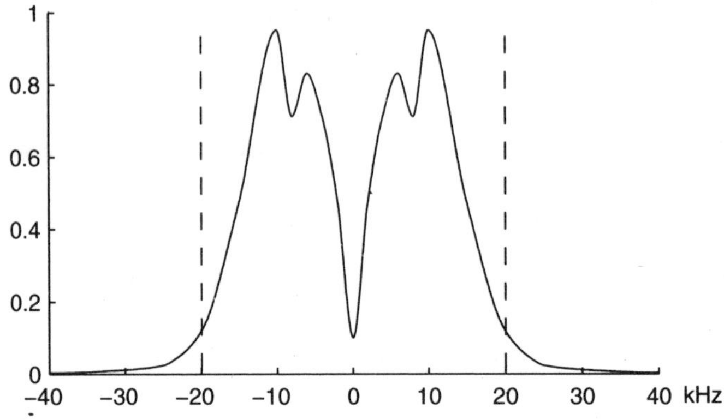

Figure 5.6. *Fourier transform $\widehat{s}(f)$ of a sound with a low frequency density outside the interval [−20 kHz, 20 kHz]*

Figure 5.6 is a typical representation of what the Fourier transform of such a sound could be: there are a few high frequencies left outside the dashed lines located at both ends of the [−20 kHz, 20 kHz] band[2]. Let us assume that this sound is sampled at a frequency of 40 kHz. An aliasing similar to what was described in the previous example will then also occur in this case: it will take effect with respect to each of the vertical axes with x-coordinates $-F_e/2 = -20$ kHz and $F_e/2 = 20$ kHz. The two aliasings are shown in dashed lines on the left graph[3], Figure 5.7. The result of the aliasing will overlap with the initial sound and distort it slightly.

What will actually be heard after the signal has gone through the DAC is the sum of the original sound and of the aliasing, shown in a dashed-dotted line in the graph on the right of Figure 5.7. In the high frequency part of the interval [−20 kHz, 20 kHz], the resulting sound has slightly more depth because of the contributions above 20 kHz from the original sound. The lower the values of $|\widehat{s}(f)|$ outside the interval [−20 kHz, 20 kHz], the less noticeable this distortion becomes. This is why sampling is usually done after the recorded sound has been filtered with a low-pass filter. That way, there is no risk of having ultrasounds that went unnoticed become an unwanted hiss, like the one seen above, which could have a disastrous effect when you listen to your favorite Mozart *andante*!

2. But we will neglect the frequencies outside the [−40 kHz, 40 kHz] band, because that would require theoretical developments beyond the scope of this book.

3. This is actually only valid for the even part of $\widehat{s}(t)$, that is, its real part, shown on this graph. For the odd part (the imaginary part), the aliasing comes with a change of sign, the same thing that would have occurred if what we had analyzed previously was a sine instead of a cosine.

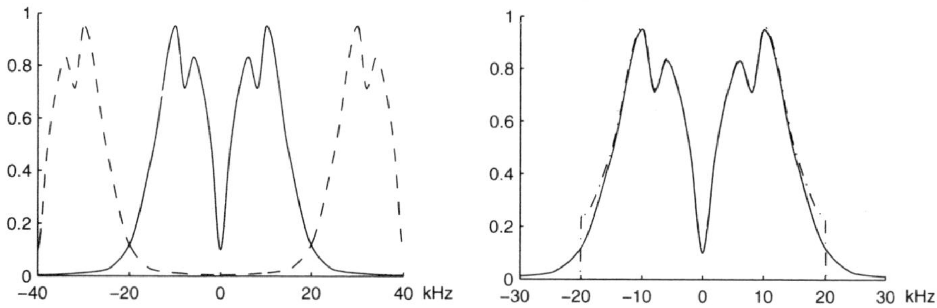

Figure 5.7. *Left: $\widehat{s}(f)$ in a full line and the two aliasings in dashed lines. Right: $\widehat{s}(f)$ in a full line and, in a dashed-dotted line, the sum of the three spectra, limited to the interval [−20 kHz, 20 kHz]. This is the sound that will come out of the DAC, with the high frequency part slightly altered.*

Finally, here is in Figure 5.8 the spectrogram of the ascending sound

$$s(t) = \cos(2\pi(1\,000t + 2\,000t^2))$$

the *instantaneous frequency* of which progressively increases with time, and is given by the formula

$$f_{\text{inst}}(t) = \frac{d}{dt}(1\,000t + 2\,000t^2) = 1\,000 + 4\,000t. \tag{5.5}$$

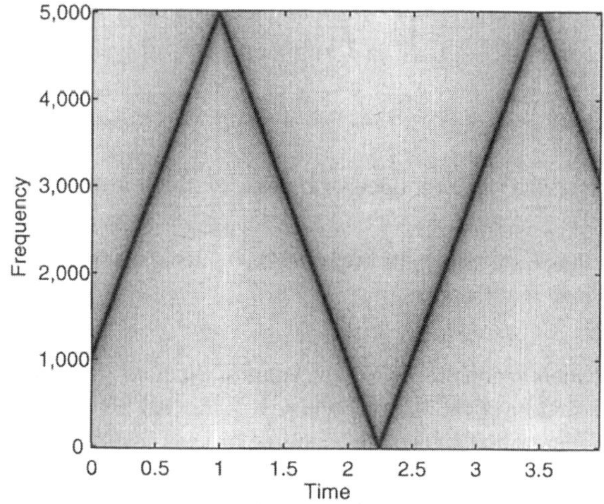

Figure 5.8. *Ascending sound and aliasing effect: are you going up or down?*

What we have here is a synthetic sound, and the samples were produced using MATLAB by calculating the values $s(t_n)$ with $t_n = n/F_e$ and $F_e = 10{,}000$. Can you explain what happens in particular at the instant $t = 1$ s?

5.1.2. *Quantization*

Faced with the practical impossibility of storing the real numbers $s_n = s(n\tau)$, which can have an unlimited number of decimals and can assume an infinity of distinct values, we have to resort to quantization, which consists of converting (encoding) each real number as a *word* with finite length, taken itself from a finite set, and which represents an *approximation* of this real number. The *quantum* in question is the minimum gap needed between two numbers so that they are coded differently. This conversion comes of course with an irreparable loss of information, a second obstacle to the perfect reconstruction of the original sound. Here, we will be describing the simplest quantization process: *uniform quantization* [MOR 95].

Consider N samples s_n, $n = 0, 1, 2, ..., N-1$, that we wish to encode as binary numbers (the words) using b bits. For example, in the case $b = 3$, these words are 000, 001, 010, 011, 100, 101, 110 and 111, the base 2 notations for the integers $0, 1, 2, ..., 7$. In all, there are $L = 2^b$ of them. The standard case $b = 16$ leads to 65,536 words. We then choose a value of A such that

$$-A \leq s_n < A \text{ for } 0 \leq n \leq N-1.$$

With the L words at our disposal, *uniform* quantization consists of:

– dividing the interval $[-A, A[$ in L subintervals I_k of *equal length* $h = 2A/L$:

$$I_k = [-A + (k-1)h, -A + kh[, \quad k = 1, 2, \ldots, L,$$

where the choice of the interval, open or closed, on the left or on the right, is rather arbitrary;

– assigning to each sample s_n its binary code b_n, which is the coded base 2 number of the interval where it is located.

For the subsequent reconstruction, each value s_n (actually each b_n) is also assigned a number r_n, the middle of the interval where it is located. Figure 5.9 (which uses the same example as Figure 5.2) shows the resulting process for 3-bit encoding. Notice by the way that the representation of certain samples is not very satisfactory. In the present case, a non-uniform quantization, with smaller intervals near 0, would probably have given a better result. Table 5.1 gives the associated values of b_n and r_n for the seven first samples s_n.

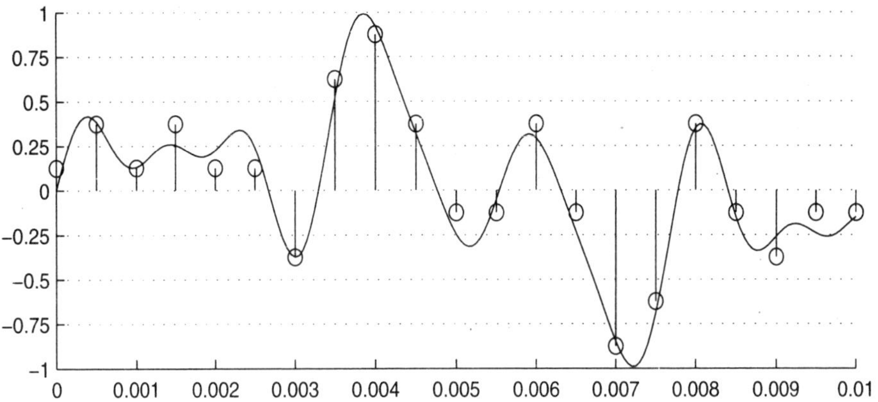

Figure 5.9. *3-bit sampling and quantization. There are 8 intervals in this case. The value r_n assigned to a sample is the middle of the interval where it is located, and its binary code is the number of the interval*

s_n	0.000	0.386	0.131	0.255	0.224	0.241	−0.373
r_n	0.125	0.375	0.125	0.375	0.125	0.125	−0.375
b_n	100	101	100	101	100	100	010

Table 5.1. *Quantization results for the samples s_n: digital approximation r_n and encoding b_n*

5.1.2.1. *Error due to quantization*

For each b-bit encoded sample, the error or *quantization noise* $\varepsilon_n = s_n - r_n$ is such, by definition, that

$$|\varepsilon_n| \leq \frac{h}{2} = \frac{A}{L} = \frac{A}{2^b}.$$

The *signal-to-noise ratio* SNR is the measurement in dB of the ratio between the sound intensity (see section 1.1.4) of the signal I_s and its noise I_b, hence, if we choose for I_s its maximum value $I_s = A^2$ and $I_b = \varepsilon_n^2$:

$$SNR = 10\log\frac{I_s}{I_b} \geq 10\log\frac{A^2}{(A/2^b)^2} = 20\,b\log 2 \simeq 6\,b \text{ dB}.$$

In particular, *adding a bit increases the signal-to-noise ratio by 6 dB*. For a quality referred to as *voice grade*, a $b = 12$-bit quantization is used, which leads to an SNR of 72 dB. For audio quality with a 16-bit quantization, the SNR increases to 96 dB, which makes the quantization noise virtually undetectable (see Chapter 4). The same estimate is obtained (but more rigorously) for the SNR by assuming that the values of the signal s_n have a uniform probability over the interval $[-A, A]$, and that the values of the error ε_n have a uniform probability over the interval $[-h/2, h/2]$. In that case, the respective average intensities of s_n and ε_n (that is, their quadratic mean) are given by

$$\overline{I}_s = \int_{-A}^{A} x^2 \frac{dx}{2A} = \frac{A^2}{3},$$

$$\overline{I}_b = \int_{-h/2}^{h/2} x^2 \frac{dx}{h} = \frac{h^2}{12} = \frac{1}{12}\left(\frac{2A}{2^b}\right)^2,$$

and we end up with the same value as before

$$SNR = 10 \log \frac{\overline{I}_s}{\overline{I}_b} = 10 \log 2^{2b} \simeq 6\, b \text{ dB}. \tag{5.6}$$

Of course, this value will not be the same when the distribution is no longer uniform, precisely the case where it may be appropriate to use a quantization that is also non-uniform.

5.1.3. *Reconstruction of the sound signal*

The reconstruction of the sound signal is based on the Shannon theorem, which states that if its hypotheses are met, we have

$$s(t) = (h * s_e)(t)$$

where $h(t)$ is the ideal low-pass filter (1.26) with cut-off frequency $F_e/2$ and $s_e(t)$ is the sampled signal (5.1). But here, we no longer have the samples s_n, only values approximated by quantization r_n. Therefore, the reconstructed signal will be

$$r(t) = (h * r_e)(t) \simeq s(t)$$

where $s_e(t)$ has been replaced by its approximation

$$r_e(t) = \tau \sum_{n=-\infty}^{+\infty} r_n\, \delta_{n\tau}(t). \tag{5.7}$$

Then comes along a third obstacle to perfect reconstruction: we do not know how to generate a 'train' of Dirac impulses $\delta_{n\tau}$. At this point, we are at the sound card level (the DAC), the task of which is to produce these impulses and to filter their sum. All we are able to do is create brief, more or less rectangular impulses, such as the ones shown in Figure 5.3. This will add a final distortion factor, that can partially be corrected using the appropriate filter. This makes the design of sound cards something of an art!

5.2. Audio compression

Digital sound compression is based on the idea of keeping *only what can be heard*, and therefore, relies on the psychoacoustic properties discussed in Chapter 4. It comes after the sampling and quantization procedures we have just described. Here, we will be pointing out the main ideas involved in compression algorithms, the most famous of which today is the MPEG Layer-3 encoding, otherwise known as MP3. The encoding comprises two distinct steps:

– a first step, *psychoacoustic compression*, that entails the loss of some information: the information considered useless in regards to the properties of hearing;

– a second step, *entropy compression*, that performs a lossless compression of the previous step's result.

The complete description of these two steps is rather technical (see for example [MOR 95]), thus we will only outline the main points. Furthermore, the first step, as it is implemented by MP3, resorts to the use of *filter banks*, used in signal techniques, and which are beyond the scope of this book (however, see Study problem 5.4.4). Therefore, we will only be presenting a modified version, but based on the same idea.

5.2.1. *Psychoacoustic compression*

In a nutshell, psychoacoustic compression consists of switching over to the frequency domain, in order to perform a *quantization of the Fourier components* using a *variable* number of bits that takes into account the properties of hearing.

Before we go into the details, we can make the following observation, to shed light on why it is useful to switch over to the frequency domain for musical sounds: let us assume for example that we have a periodic sound, sampled at a frequency F_e and 16-bit quantized. Based on (5.6), the SNR is equal to 96 dB. As additional time compression, we could for example use only 8 bits to encode these samples. The SNR would then drop to 48 dB, and the quantization noise would become noticeable (see `AM` website): what we would hear is the initial sound, but along with a slight hiss (white noise). On the other hand, if the Fourier coefficients are 8-bit quantized, there will be a slight modification to the relative weights of the different harmonics, but no additional frequencies will be added. Therefore, there will be no hiss in the reconstructed sound, only a modification of the timbre, hardly, if not at all, noticeable at this level of quantization. We can even go down to 4 bits and still the produced distortion would not be too bad, whereas the noise produced by a 4-bit time quantization would be downright unbearable!

Now for the details. Our starting point, which will serve as an illustration, is a sound sampled at the frequency F_e, 16-bit quantized and denoted by $r_e(t)$. In order to process it, we start by cutting it up into small segments each containing 512 samples. Each segment is obtained by multiplying the signal by a sliding window w, like the one we used in section 1.4.2, and such that two consecutive segments overlap slightly. Figure 5.10 shows how the second segment is obtained.

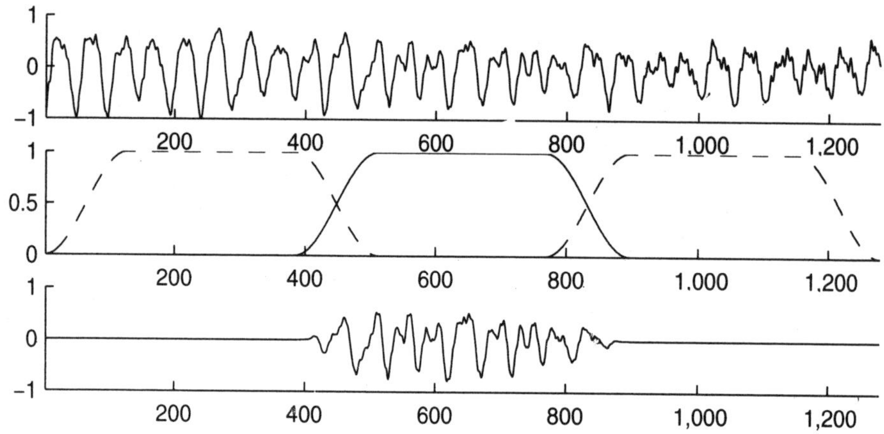

Figure 5.10. *At the top, the sampled signal r_e (shown in a full line for clarity), an excerpt of our beloved tune* Norma. *In the middle, three consecutive windows w with a full line for the one currently being used. At the bottom, the segment with the $u_n = w_n \times r_{n+m}$ that are to be analyzed and compressed, with in this case $m = 3 \times 512/4 = 384$, and $n = 0, 1, \ldots, 511$.*

This segment u currently being analyzed is comprised of 512 16-bit encoded samples u_n that we wish to compress. The psychoacoustic properties we are going to rely on have to do with the frequency components found in the signal, hence the first operation will consist of calculating its DFT (discrete Fourier transform). The coefficients of this DFT are the 512 complex coefficients $\widetilde{c}_n$ given by formula (1.16), which we will denote here by c_n to simplify. The corresponding intensity

$$L_I(n) = 20 \log |c_n|$$

is shown in a full line in Figure 5.11. As in Figure 1.25, the x-coordinates do not represent the index n, but instead the associated frequency $f_n = n/T$, where $T = 512/F_e$ is the duration of the segment being analyzed.

In particular, some peaks in frequency show up, indicated by small circles. When these peaks have an intensity located at least 7 dB above their close neighbors, they are referred to as *tonal components*. The other peaks are referred to as *non-tonal* components. With MP3 encoding, a differential process is applied to these two types of components, but we will not describe it here.

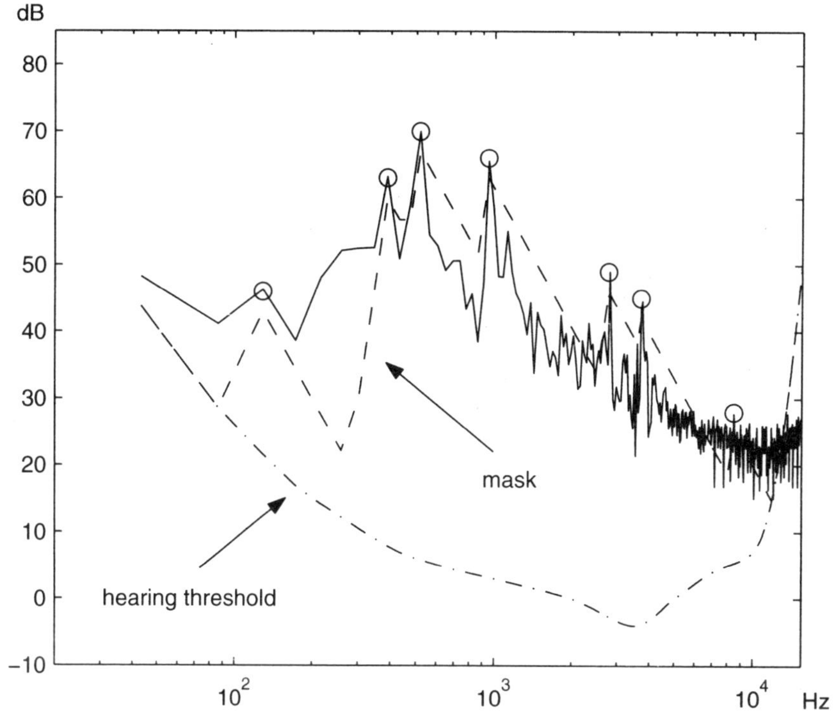

Figure 5.11. *Spectrum (the DFT) of the 512 samples u_n and global mask of the spectrum. This mask takes into account both the masking effect of the frequency components with strong intensities and the hearing threshold (3 phon equal-loudness)*

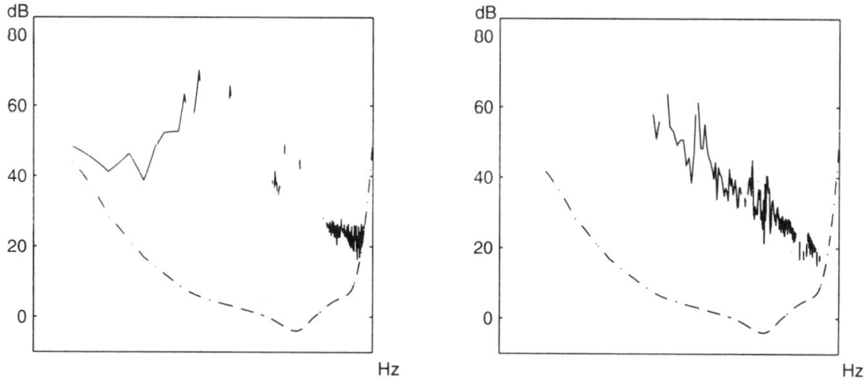

Figure 5.12. *Separating the parts of the spectrum located above (figure on the left) and below (figure on the right) the mask, with the intent of performing differential compression (variable number of bits depending on the Fourier coefficient)*

The same figure shows in dashed lines the frequency masks produced by these peaks, those we had described in section 4.4. In this case, we have 7 peaks and 7 associated masks $M^{(i)}$, $i = 1, 2, ..., 7$. The mask M shown in the figure is the *upper envelope* of these masks and of the hearing threshold S, meaning that its discrete values M_k are defined by

$$M_k = \max\left\{M_k^{(1)},\ M_k^{(2)},\ \ldots,\ M_k^{(7)},\ S_k\right\}, \quad k = 0, 1, 2, \ldots$$

Thus, the part of the DFT that is significant to our hearing is the one located *above* the mask M, shown on the left of Figure 5.12. Let J be the set of indices of the corresponding coefficients c_n. The part located *below* the mask, shown on the right of Figure 5.12, should be inaudible since it is masked (covered) by the first part. The set of indices of the coefficients c_n relevant to this second part is denoted by K.

Psychoacoustic compression then consists of *quantizing the DFT's coefficients, using less bits to encode the 'inaudible' coefficients c_k, $k \in K$, than to encode the 'audible' coefficients c_j, $j \in J$*. Furthermore, because of the formula that gives us the c_n:

$$c_n = \frac{1}{N} \sum_{k=0}^{N-1} U_k \exp(-2i\pi nk/N),$$

we notice that we have a conjugated symmetry, $c_{N-n} = \overline{c_n}$, and therefore, we simply have to know half of the coefficients plus one[4]: c_0, c_1, ..., c_{256}. Thus these first 257 complex coefficients (512 real numbers because c_0 and c_{256} are real) will be the ones that are quantized, and stored after the entropy compression described in section 5.2.2.

During the subsequent *decompression*, usually while the sound is being played, all of the 512 coefficients c_n are recalculated, their inverse DFT (1.17) is performed, and finally the successive segments are added to one another. This would result in exactly the samples r_n had we not encoded the c_n on a limited number of bits. The initial sound is thus reconstructed with an error that theoretically should only have an effect on the components that cannot be perceived.

An excerpt of the initial sound and its reconstruction obtained through this process are shown in Figure 5.13. The initial sound sample was comprised of 512 16-bit encoded values, in all 8,704 bits. Once it has been encoded and compressed using entropy encoding, which will be discussed later on, the signal's DFT only takes up 2,280 bits, hence a compression rate of 74 %, with a relative mean square error

4. A different version consists of conducting all of the calculations with real numbers by extending the samples u_n in an even way. This version is called the Discrete Cosine Transform (DCT), since an even function is expanded as a series of cosines. It is actually the most frequently used.

of 20 %. With some (a lot) refinements to this process, the MP3 standard achieves a compression rate of about 90 % while maintaining excellent sound quality.

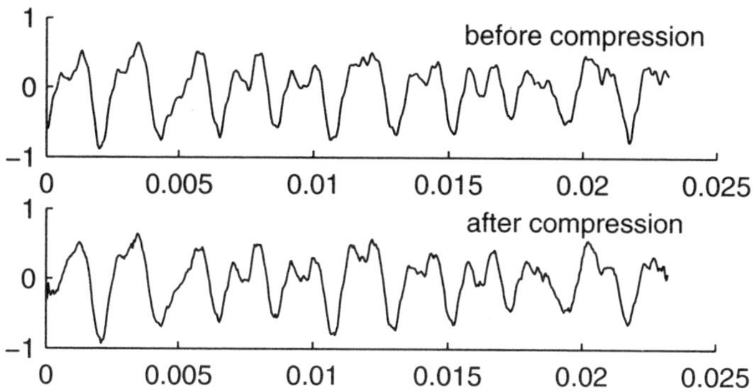

Figure 5.13. *Initial sound (top) and sound after compression-decompression (bottom). During the psychoacoustic compression, the part of the DFT located above the mask was 8-bit encoded, whereas the part located below was 2-bit encoded*

5.2.2. *Entropy compression*

Following the psychoacoustic compression, we are left with a sequence of binary numbers or words m_1, m_2, ... m_k, ..., of variable lengths. These are the DFT's quantized coefficients. It often occurs for two consecutive words to be identical (particularly zeros), thus we begin by replacing each sequence of a word m_k repeated p times with the two words $p\ m_k$. The result is a second, shorter sequence of binary numbers. These words, which form a text, are themselves part of a *dictionary* comprising a finite number of words d_i.

We then proceed to a *change of dictionary*, a bit like translating the text in a new language, based on the following, obvious and remarkably simple idea: *encoding the most frequent words as short words*. This is done by counting the number of occurrences in the text for each word d_i, and, depending on the *probability* (number of occurrences/total number) of each of these words, a new dictionary is created based on this idea. Two examples of such dictionaries are given in Table 5.2.

	word1	word2	word3	word4	word5	word6	word7	word8
dictionary1	0	10	110	1110	11110	111110	1111110	11111110
dictionary2	00	01	10	1100	1101	1110	111100	111101

Table 5.2. *Two possible encodings of eight words arranged in decreasing order of occurrence frequency*

Once they are translated, words are written one after the other, without any spaces in between. Therefore, for the resulting text to be *decodable*, no word must be the beginning of another word, as it would make several different interpretations possible. For example, if the three words A, B and C are encoded as 0, 1 and 10, the text 10 could be interpreted both as C and as BA. This is referred to as the *prefix condition*.

Several strategies exist to build this new dictionary, implemented for example in applications such as *winzip*, or in `unix` commands such as *gzip* or *compress*. Particularly, one of them is optimal: the Huffman algorithm (see for example [MOR 95]). If we consider the same example as before, it produces the dictionary in Table 5.3. Of course, when you send a text that was translated this way, do not forget to send the dictionary!

	word1	word2	word3	word4	word5	word6	word7	word8
probability	0.25	0.22	0.19	0.11	0.11	0.05	0.05	0.02
Huffman encoding	10	00	110	010	011	1110	11110	11111

Table 5.3. *Huffman's optimal encoding for a given probability*

5.3. Digital filtering and the Z-transform

We end this chapter with a few elements of *digital filtering* that will be useful to us in Chapter 6. For analog signals, we saw in section 1.5.2 two fundamental relations (1.25) and (1.24) on filtering: if a signal x is passed through a filter with impulse response h (hence a transfer function $\widehat{h}$), the output signal y is such that

$$y(t) = (h * x)(t),$$

$$\widehat{y}(f) = \widehat{h}(f)\,\widehat{x}(f).$$

We are now going to see what these relations mean for the samples.

5.3.1. *Digital filtering*

Let x_n, h_n and y_n be the samples of the analog signals x, h and $y = h * x$, respectively. We will admit the following result.

If Shannon's condition is verified by x or h: at least one of the two is $[-B, B]$ band-limited with $F_e > 2B$, then y is $[-B, B]$ band-limited and we have

$$y_n = \frac{1}{F_e} \sum_{k=-\infty}^{+\infty} h_{n-k} x_k. \tag{5.8}$$

This relation defines what is referred to as the *discrete convolution* of the signals $(h_n)_{n\in\mathbb{Z}}$ and $(x_n)_{n\in\mathbb{Z}}$, and we keep the same notation by writing[5]

$$y = h * x.$$

Notice by the way the strong analogy with continuous convolution (1.23) given here as a reminder:

$$y(t) = (h * x)(t) = \int_{-\infty}^{+\infty} h(t-u)x(u)\, du.$$

The integral has simply been replaced with a sum, and t and u have been replaced with n and k, respectively.

What is important to note is that *any analog filtering*, which relies on acoustic or electronics devices, depending on the case, *can be performed in an equivalent way* (more or less) *digitally*, so long as the Shannon condition is verified (more or less) and the samples of the filter's impulse response are available.

When the number of non-zero coefficients h_n is *finite*, the filter has a *finite impulse response*, and it is referred to as an FIR filter. Otherwise, it has an *infinite impulse response*, and it is referred to as an IIR filter.

5.3.2. *The Z-transform*

What happens to the sampled signals in the frequency domain? These signals have a Fourier transform, but its definition requires a mathematical tool beyond the scope of this book: distribution theory. We will be describing another tool quite similar to the Fourier transform: the *Z-transform*. As we are going to see, it can be used very conveniently to represent operations that have to do with discrete filtering.

5.3.2.1. *Definition*

Let $x = (x_n)_{n\in\mathbb{Z}}$ be a discrete signal. The Z-transform of the signal x, denoted by X, is the function of the complex variable z such that

$$X(z) = \sum_{n=-\infty}^{+\infty} x_n z^{-n}. \tag{5.9}$$

This sum is usually not defined for any z, only for a part of the complex plane, with the following expression, called the *region of convergence* (or ROC):

$$C(\rho, R) = \{z \in \mathbb{C}\, ;\, \rho < |z| < R\},$$

5. We use the same notation as for the analog signal x and the discrete signal represented by the sequence of samples x_n. This may lead to some confusion, but the context should usually help determine which is the signal in question.

over which the function X is *holomorphic* (which means differentiable with respect to z).

In practice, it is always possible to make modifications and end up with the case where $x_n = 0$ for any $n < 0$. Such a discrete signal is said to be *causal*. In that case, we have $R = +\infty$ and if $\rho < \infty$, $X(z)$ has a limit in $+\infty$, the number x_0, which is what we will assume. The question that comes to mind is this: given a function $u(z)$ that verifies these hypotheses, can we associate it with a discrete signal x such that $X(z) = u(z)$ in the region of convergence $C(\rho, \infty)$?

To answer this, consider the function $v(z) = u(z^{-1})$, defined for $|z| < 1/\rho$. According to the theory of holomorphic functions, it has a series expansion of the form

$$v(z) = \sum_{n=0}^{+\infty} x_n z^n, \quad x_n \in \mathbb{C}.$$

We then have

$$u(z) = v(z^{-1}) = \sum_{n=0}^{+\infty} x_n z^{-n}. \tag{5.10}$$

Thus we have found a discrete signal $x = (x_n)_{n \in \mathbb{N}}$ ($\mathbb{N}$ being the set of natural numbers) such that $X(z) = u(z)$. Additionally, because the series expansion of v is unique, this signal x is the *unique causal signal* with $u(z)$ as its Z-transform.

5.3.2.2. *Effect of a delay*

If a sampled signal x is delayed by m samples, which is equivalent to a time delay for the analog signal of $r = m/F_e$ s, we get a discrete signal y with the coefficients

$$y_n = x_{n-m}.$$

The Z-transform of y is written

$$Y(z) = \sum_{n=-\infty}^{+\infty} y_n z^{-n} = \sum_{n=-\infty}^{+\infty} x_{n-m} z^{-n} = \sum_{n=-\infty}^{+\infty} x_n z^{-n-m} = z^{-m} \sum_{n=-\infty}^{+\infty} x_n z^{-n},$$

which in the end leads to

$$Y(z) = z^{-m} X(z). \tag{5.11}$$

Therefore, the m-sample delay simply amounts to multiplying the Z-transform by z^{-m}.

5.3.2.3. *Filtering and Z-transform*

To finish this, we will need the two following important results to deal with and interpret discrete filtering.

Let x and h be two discrete signals, and let $y = h * x$ be the convolution product defined by (5.8). Let X, H and Y be their respective Z-transforms, defined on regions of convergence denoted by C_X, C_H and C_Y. Then, for any $z \in C_X$ and C_H, we have $z \in C_Y$ and

$$Y(z) = H(z)X(z). \tag{5.12}$$

Again, notice the analogy with the relation $\widehat{y}(f) = \widehat{h}(f)\widehat{x}(f)$ which applies to the associated analog signals.

Also, given the equivalence with the continuous filtering described in section 5.3.1, and by assuming the hypothesis that the signal x is sampled at the frequency F_e of an analog sound that verifies the Shannon condition, it is useful to be able to say which transfer function is associated with the discrete filtering $y = h * x$. So as not to confuse it with the Z-transform of h, this transfer function will be denoted by $H_t(f)$. This function is then related to the Z-transform of h by the following formula:

$$H_t(f) = H(\exp(2i\pi f/F_e)). \tag{5.13}$$

In other words, the transfer function's values are obtained by taking the values of the Z-transform of h *on the unit circle* of the complex plane.

The equivalence with continuous filtering means the following: after reconverting the digital sound $y = h * x$ back to an analog sound by passing it through the DAC, still denoted by y, its Fourier transform will be

$$\widehat{y}(f) = H_t(f)\widehat{x}(f),$$

where x now refers to the analog sound that had produced the samples x_n. This makes it possible to interpret the effect of a digital filtering by considering the values of $H_t(f)$.

Under the Shannon condition, we have $\widehat{x}(f) = 0$ if $|f| \geq F_e/2$. Therefore, the only values of $H_t(f)$ that concern us are those for which $f \in [-F_e/2, F_e/2]$, or $2f/F_e \in [-1, 1]$. Additionally, by replacing x with h and by choosing $z = \exp(2i\pi f/F_e)$ in (5.9), we have

$$H(\exp(2i\pi f/F_e)) = \sum_{n=-\infty}^{+\infty} h_n \exp(-2ni\pi f/F_e),$$

hence we infer, using (5.13), that for a filter with real coefficients h_n we have

$$H_t(-f) = \overline{H_t(f)},$$

and therefore, we simply need to know the values of $H_t(f)$ for $f \geq 0$. In the end, the values that are to be considered are the $H_t(f)$ for which

$$\frac{2f}{F_e} \in [0,\ 1],$$

which is the way discrete filters are represented in the software MATLAB. The four basic filter models are shown in Figure 5.14.

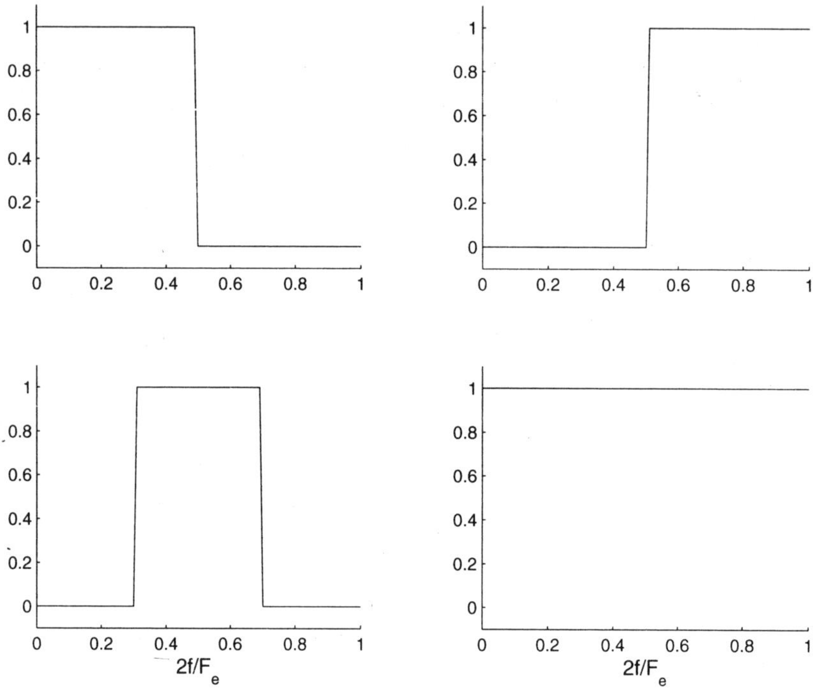

Figure 5.14. $|H_t(f)|$ *plotted for the four most common filters. The phase of* $H_t(f)$ *may vary depending on* f *(not shown). Top left: low-pass filter with cut-off frequency* $0.5 \times F_e/2$, *and on the right: high-pass filter with cut-off frequency* $0.5 \times F_e/2$. *Bottom left: band-pass filter with cut-off frequencies* $0.3 \times F_e/2$ *and* $0.7 \times F_e/2$, *and on the right: all-pass filter.*

5.4. Study problems

5.4.1. *Nyquist criterion (*)*

In this chapter, we analyzed what happened when the Nyquist criterion was not met, taking as our example the sound $s(t) = 2\cos(2\pi ft)$, with $f = 27$ kHz and $F_e = 30$ kHz. Conduct the same analysis for the sound $s(t) = \sin(2\pi ft)$.

5.4.2. *Aliasing of an ascending sound (*)*

Plot according to f the effective frequency of the sound reconstructed after sampling the sound $s(t) = \cos(2\pi f t)$ at the frequency $F_e = 1{,}000$ Hz. Based on this result, interpret Figure 5.8.

5.4.3. *Another example of reconstruction (***)*

The Shannon theorem states that if the Nyquist criterion is met, then the reconstruction of the sound $s(t)$ by an ideal low-pass filtering of the sampled sound

$$s_e(t) = \tau \sum_{n=-\infty}^{+\infty} s(n\tau)\,\delta_{n\tau}$$

is exact:

$$s(t) = (h * s_e)(t).$$

We are going to replace each Dirac impulse $\delta_{n\tau}$ with the approximation discussed in section 5.1:

$$\delta_{n\tau}(t) \simeq u_\varepsilon(t - n\tau)$$

with

$$u_\varepsilon(t) = \begin{cases} 1/\varepsilon \text{ if } |t| < \varepsilon/2, \\ 0 \text{ otherwise,} \end{cases}$$

where we assume that $0 < \varepsilon < \tau$, τ being the sampling period.

1) By making the variable change $t' = (t - n\tau)/\varepsilon$, check that for any function φ continuous on $\mathbb{R}$, we have

$$\lim_{\varepsilon \to 0} \int_{-\infty}^{+\infty} u_\varepsilon(t - n\tau)\varphi(t)\,dt = \varphi(n\tau) = \int_{-\infty}^{+\infty} \varphi(t)\delta_{n\tau}(t)\,dt.$$

2) The sampled sound $s_e(t)$ is replaced with

$$s_\varepsilon(t) = \tau \sum_{n=-\infty}^{+\infty} s(t) u_\varepsilon(t - n\tau). \tag{5.14}$$

Let

$$v_\varepsilon(t) = \tau \sum_{n=-\infty}^{+\infty} u_\varepsilon(t - n\tau).$$

Check that the function v_ε is τ-periodic.

3) Therefore the function v_ε can be written in the form

$$v_\varepsilon(t) = \sum_{k=-\infty}^{+\infty} c_k \exp(2i\pi kt/\tau)$$

where the c_k are its Fourier coefficients. Show that

$$\widehat{s}_\varepsilon(f) = \sum_{k=-\infty}^{+\infty} c_k \widehat{s}(f - k/\tau)$$

(we will assume that we can do the switch $\int \sum = \sum \int$).

4) The Nyquist criterion being met, we have $\widehat{s}(f) = 0$ if $|f| > B$ with $B < 1/(2\tau)$. Make a sketch over the interval $[-2/\tau, 2/\tau]$ of the plot of the function $\widehat{s}_\varepsilon$ according to $\widehat{s}$. Check that $c_0 = 1$, and compare $\widehat{s}(f)$ with $\widehat{h}(f)\widehat{s}_\varepsilon(f)$, h being the ideal low-pass filter with cut-off frequency $1/(2\tau)$. Based on this result, find the reconstruction of s based on s_ε:

$$s(t) = (h * s_\varepsilon)(t).$$

COMMENT.– In reality, $s(t)$ is replaced in (5.14) with $s(n\tau)$, which causes a small error, that decreases with ε.

5.4.4. *Elementary filter bank (**)*

Given a discrete signal $x = (x_n)_{0 \leq n \leq N}$, consider two discrete filters g and h defined by their outputs y and z, respectively:

$$y_n = \frac{1}{2}(x_n + x_{n-1}),$$

$$z_n = \frac{1}{2}(x_n - x_{n-1}),$$

with the convention that $x_n = 0$ if $n < 0$ or $n > N$. Thus the result is $y = g * x$ and $z = h * x$.

1) Calculate the coefficients g_n and h_n of these two filters.

2) Calculate their Z-transforms, and plot the modulus of their transfer functions $|G_t(f)|$ and $|H_t(f)|$. What is the nature of each of these filters: low-pass or high-pass?

3) The outputs y and z are thus the low frequency and high frequency outputs, respectively, of the signal x. Let us assume that we wish to use this data to compress the signal. We are off on a bad start, to the extent that we have roughly multiplied by two the volume of this data! Therefore only even terms are kept: let

$$u_n = y_{2n}, \quad v_n = z_{2n}, \quad 0 \leq 2n \leq N + 1,$$

a process called *decimation*. These two signals u and v are then compressed using whatever process, but with no additional loss of information, which will not be discussed here. During the subsequent decompression phase, the goal is to reconstruct the signal we started with, x. This is done by refiltering the decompressed signals u and v with g and h in the following way: by calculating

$$p = 2g * r,$$

$$q = -2h * s,$$

where r and s are the signals referred to as *interpolated* signals

$$r = (u_0, 0, u_1, 0, u_2, 0, \ldots),$$

$$s = (v_0, 0, v_1, 0, v_2, 0, \ldots).$$

Show that $p_{n+1} + q_{n+1} = x_n$ for $0 \leq n \leq N$. Thus, despite appearances, no information was lost and x could be reconstructed. This whole sequence of operations:

filtering $\longrightarrow$ decimation $\longrightarrow$ interpolation $\longrightarrow$ filtering $\longrightarrow$ addition

forms what is called a *perfect reconstruction filter bank*.

5.5. Practical computer applications

5.5.1. *Spectrum aliasing*

In a command file, program the sampled sound obtained from the sound

$$s(t) = \cos(2\pi(1\,000\,t + 2\,000\,t^2))$$

using $F_e = 11{,}025$ Hz as the sampling frequency and with a duration of a few seconds. What would we hear if we listened to the continuous sound (calculate the instantaneous frequency)? Listen to the resulting digital sound; what can you notice? Visual confirmation: use the function `specgram` to analyze the sound.

5.5.2. *Quantization noise*

The effect of the following instruction: `x4 = (round(2^3*x+.5)-.5)/2^3;` is to simulate 4-bit quantization of the sound contained in the vector `x` if this vector has all of its values lying strictly between -1 and 1. Can you explain why? Write a program that quantizes the samples of a sound composed of a few harmonics, consecutively using 8, 6 and 4 bits and listen to the quantization noise that appears when the modified sound is played. Then compare this with what happens when the amplitudes of each harmonic are quantized.

Chapter 6

Synthesis and Sound Effects

The first to practice the synthesis of musical sounds may have been Organ manufacturers, who by combining several pipes attempted to recreate the human voice (which led to the *regal* category) or the sounds of orchestra instruments (see section 2.5.1.3). For example, in *gamba stops*, two pipes play together slightly out of tune in order to produce a beat, simulating the undulation of string instruments: this is called the *voix céleste*, a typical organ stop of the romantic narrative. The 20th century saw the coming of electronic synthesis instruments (organs, various keyboards), combining, among other things, oscillating circuits and filters to produce musical sounds. Composers such as Pierre Henry took advantage of the new possibilities offered by these electro-acoustic sounds. Today, a remarkable advantage provided by digital technology is the possibility to create all kinds of sounds with a computer, from the imitation of acoustic or electronic instruments to the creation of entirely synthetic sounds, such as the *Modalys* synthesis system developed by the IRCAM (the French Acoustics and Music Research and Coordination Institute) [PRO 95]. We are going to see a few simple processes to create such sounds.

Even though it is difficult to precisely define what makes them different from an actual sound, we can say that sound effects are modifications of pre-existing sounds. They have probably always been used, and acoustic instruments provide a few examples: pedals on harpsichords and pianos, different types of mutes for the trumpet, but also the vibrato produced by the musician himself. With the advent of electronic instruments, rock musicians became avid users of sound effects: the Leslie effect in electronic organs, guitar pedals (*wah-wah*, compressor, reverb, saturation), etc. Now, digital technology can reproduce all of these effects and create an infinite number of new ones [DAF 02]. Because we are short on space, we will only be describing a few!

6.1. Synthesis of musical sounds

Digital synthesis techniques of musical sounds can be divided into two main categories: those that rely on *physical modeling*, and those that resort to predefined *signal models*.

Synthesis by physical modeling consists of starting with a simplified physical model of the instrument we are trying to produce or reproduce, and then to perform digital computations that lead to the model's response to a given excitation. A basic example of this is given in Study problem 2.6.5, which manages in particular to account for the remanence effect. We will not go into this any further, but it is a field of very active research [PRO 95], and is starting to appear in digital instruments sold to the public. The essential difficulty of physical modeling is to take into consideration the linear and non-linear (at the vibrator's level) effects both in a way simple enough to be able to conduct the computations in real-time, and subtle enough to obtain a good sounding result – two essential requirements for musicians.

In synthesis based on predefined signal models, valid both for electronic and digital applications, there are at least four types of techniques, which can actually be combined:

– *subtractive* synthesis, where we start with a sound that has a lot of depth, and the spectrum of which is modeled by filtering it to obtain the desired sound (like the sculptor who starts with a full block from which he removes material);

– *additive* synthesis, where we proceed by adding pure sounds with different frequencies, modifying weights and their envelopes (like the sculptor who proceeds by successive additions of material);

– *frequency modulation* (FM) synthesis invented by Chowning, inspired from the radio wave transmission technique;

– synthesis using previously *sampled sounds*.

The first two techniques can hardly render all of the 'life' of a sound produced by an acoustic instrument, that is, its complexity, its unexpected or even random variations, its 'flaws' and its oddities. Although the last technique is supposed to solve this problem since it reproduces actual sounds, it does not avoid this criticism either because it still reproduces them in the same way (when it comes to instruments sold to the public). The third technique is a bit different, because it has mainly created new sounds, that have widely been adopted, particularly in rock and jazz.

6.1.1. *Subtractive synthesis*

Subtractive synthesis requires as a starting point a sound with a lot of depth, but nonetheless with enough structure that a musical sound can be extracted from it, and works by filtering this sound. In this respect, it imitates the resonator of a musical instrument that filters the signal produced by the vibrator, or the vocal tract that filters the sound produced by the vocal chords, causing formants to appear (Figure 1.29). Take for example as the starting point a triangular (Figures 1.14 and 2.16) or rectangular (Figure 1.15) sound. It is then necessary to look for a filter with an output that imitates the sound of bowed strings in the first case, and of reed instruments in the second. This technique, however, is quite demanding computation-wise, and seems to have had less success than the others.

6.1.2. *Additive synthesis*

Using the same approach as organ manufacturers, researchers like Mathews in his modular programs Music III, IV and V (around 1960, [MAT 69, PIE 99]), followed by designers of electronic and then of digital instruments like Moog and Buchla, used additive synthesis to create synthetic sounds, imitating acoustic instruments or producing new sounds. These sounds are of the form

$$s(t) = \sum_n e_n(t, I) \sin(2\pi f_n t + \theta_n) + b(t, I), \tag{6.1}$$

which we have already come across in section 2.5.2: $e_n(t, I)$ is the envelope of the n-th harmonic or partial. The parameter I represents the intensity at which the note will be played, because we cannot just multiply everything by a same quantity to get a stronger note, since the relative intensities of the harmonics or partials can vary, as well as the envelope, as you can see from the sonogram of a flute's C5 played *crescendo*, shown in Figure 6.1.

Therefore, the weight and the shape of $e_n(t, I)$ have to vary according to the intensity I. In (6.1), a noise $b(t, I)$ was added to imitate the hiss of a wind instrument, the bow stroking a string, the percussion sound at the beginning of piano note, etc.

It thus becomes clear that when imitating an acoustic instrument, the implementation turns out to be very difficult: the sound we wish to reproduce must first be precisely analyzed, at different intensities, and a good model must be determined to modify the envelopes and the noise according to the intensity. This method is also quite demanding computation-wise.

6.1.3. *FM synthesis*

The FM synthesis technique was designed by Chowning in 1973. First, remember that there are three types of modulation used for radio wave communications: *amplitude modulation (AM), phase modulation (PM) and frequency modulation (FM)*. In all three cases, we deal with two signals:

– a *carrier* wave, which is a sinusoidal signal with a high frequency (around 1 MHz for AM, 100 MHz for FM) used as the 'transport vehicle', of the form $v(t) = \sin(2\pi f_0 t)$ (except for a phase shift);

– a *modulating* wave $m(t)$, which is the information transported by the carrier, usually a $[-B, B]$ band-limited wave with $B < f_0$.

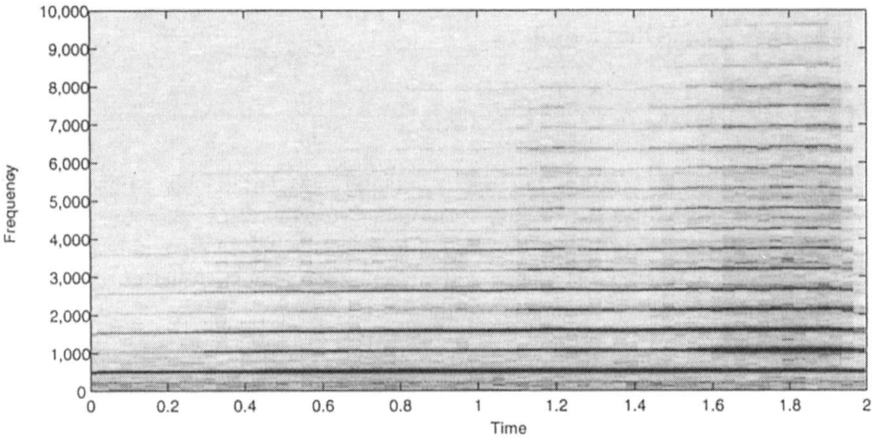

Figure 6.1. *Crescendo on a flute's C5, showing the increasing relative weights of the high harmonics*

The three modulations shown in Figure 6.2 correspond to the following signals:

$$\text{AM:}\quad s(t) = m(t)\sin(2\pi f_0 t), \quad m(t) \geq 0,$$

$$\text{PM:}\quad s(t) = \sin(2\pi f_0 t + m(t)), \quad -\pi \leq m(t) < \pi,$$

$$\text{FM:}\quad s(t) = \sin(2\pi f_0 t + 2\pi M(t)), \quad M'(t) = m(t).$$

In PM and FM modulation, it is important in the subsequent demodulation (when the sound is played) for $m(t)$ to be $[-B, B]$ band-limited with $0 < B \ll f_0$: the frequencies present in the modulating wave have to be *considerably lower* than the carrier's frequency.

Chowning's idea in 1973 best illustrates how a simple idea can revolutionize a particular field. His idea for synthesizing sounds was simply to use frequency modulation *with a modulating wave having a frequency greater than or equal to the carrier's frequency!* In other words, the sounds he created were of the form

$$s(t) = \sin(2\pi f_0 t + \beta \sin(2\pi f_1 t)/f_1)$$

with $f_1 \geq f_0$. Here, the instantaneous frequency (see (5.5)) is given by

$$f_{\text{inst}}(t) = f_0 + \beta \cos(2\pi f_1 t)$$

and $m(t) = \beta \cos(2\pi f_1 t)$. The modulating wave's frequency is f_1, and β is the *frequency excursion*: the instantaneous frequency varies inside the interval $[f_0 - \beta, f_0 + \beta]$. The result is a whole range of timbres (see AM website), that depend only on the parameters β and f_1. Figure 6.3 shows the many kinds of sounds that can be obtained, simply by modifying f_1.

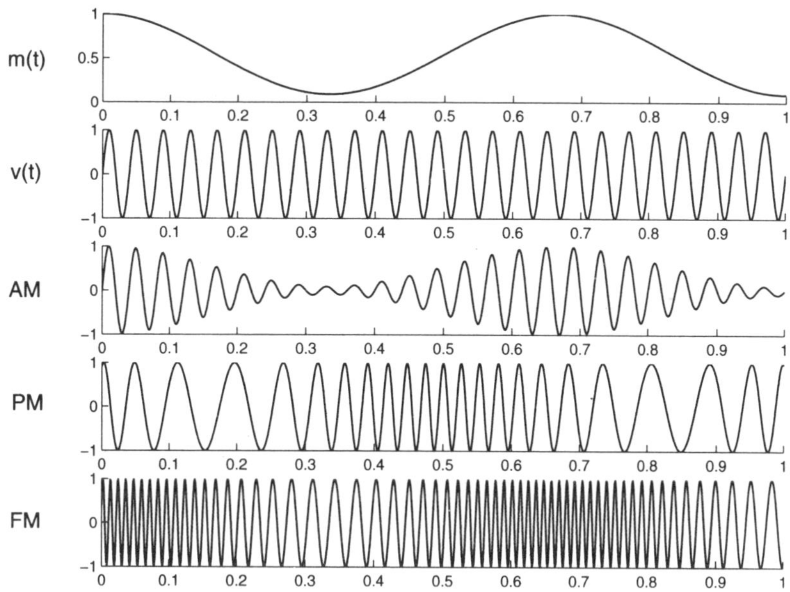

Figure 6.2. *From top to bottom: the modulating wave $m(t)$, the carrier $v(t)$ and the three types of modulation AM, PM and FM. For the PM and FM modulations, the signals $m(t)$ and $M(t)$ have been multiplied by 20 and 200, respectively, to make the variations of the signals visible.*

As for the parameter β, it makes it possible to modify the specter according to the intensity: for $\beta = 0$, we have a pure sound, that gains high harmonics as β increases (see Figure 6.4), as with most musical instruments. Thus, when digital keyboards became capable of reacting to the key's attack, or *velocity*, the timbre could be modified by increasing β at the same time as the velocity.

Additionally, this FM technique offered an easy implementation, by simply recording the carrier's sample onto a register (a memory), and then playing these samples at a variable speed. Thus was born the famous set of Yamaha DX synthesizers, including the very popular DX7. Later on, Chowning used the same technique to imitate the formants of the voice in singing, again choosing f_0 as an integer multiple of f_1, for example $f_0 = 17f_1$ or $f_0 = 25f_1$, or also by interweaving FM modulations on several levels.

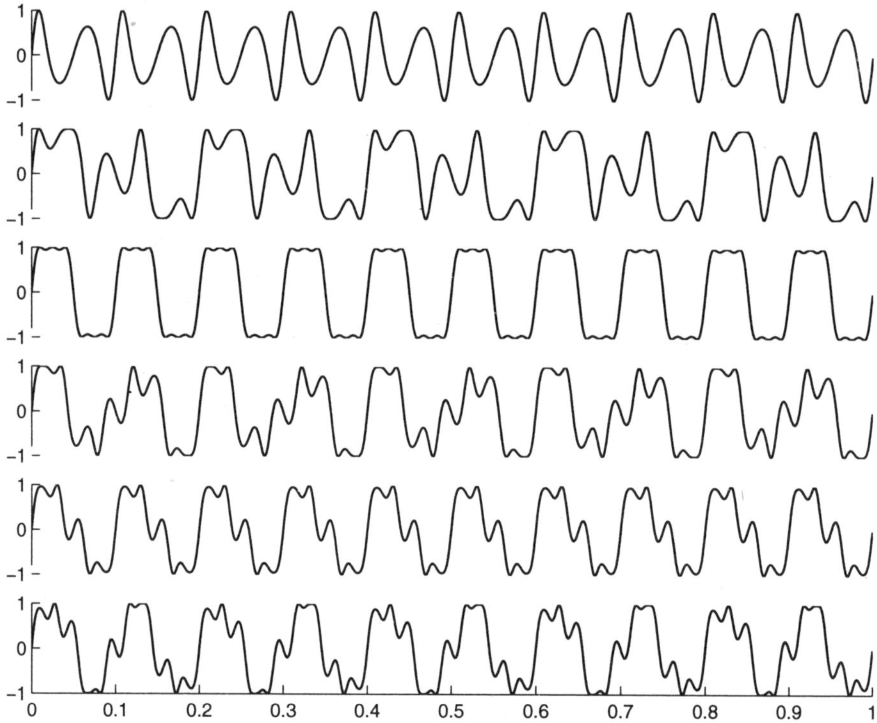

Figure 6.3. *A few examples of sounds produced with Chowning's technique. They are of the form* $\sin(2\pi f_0 t + \beta \sin(2\pi f_1 t)/f_1)$, *with* $f_0 = 10$, $\beta = 20$, *and from top to bottom:* $f_1 = 10, 15, 20, 25, 30, 35$

6.1.4. *Synthesis based on the use of sampled sounds*

Synthesis based on the use of previously sampled sounds is currently the most common technique for the imitation of acoustic (or electronic) instruments. Rather than implementing one of the syntheses we mentioned, the result of which usually is not satisfactory to musicians, pre-recorded sounds of instruments are used, and simply have to be played at the right time.

This technique requires large storage capacities. The first instruments used the same sample for several notes (one octave for example), but the sample changes were noticeable: this is because a G on a given instrument is different from a C with all the frequencies simply multiplied by 1.5, *i.e.* played 1.5 times faster. The relative weights of the harmonics also change: usually the relative weights of high harmonics decrease when the fundamental's frequency increases. With the increase in memory storage capacities, things changed to one, then several samples per note, to render the variation of a same note according to its intensity. The current trend is to conduct, based on a physical model and a set of samples for each note, a processing of these samples to build the sound according to parameters such as intensity, duration, etc.

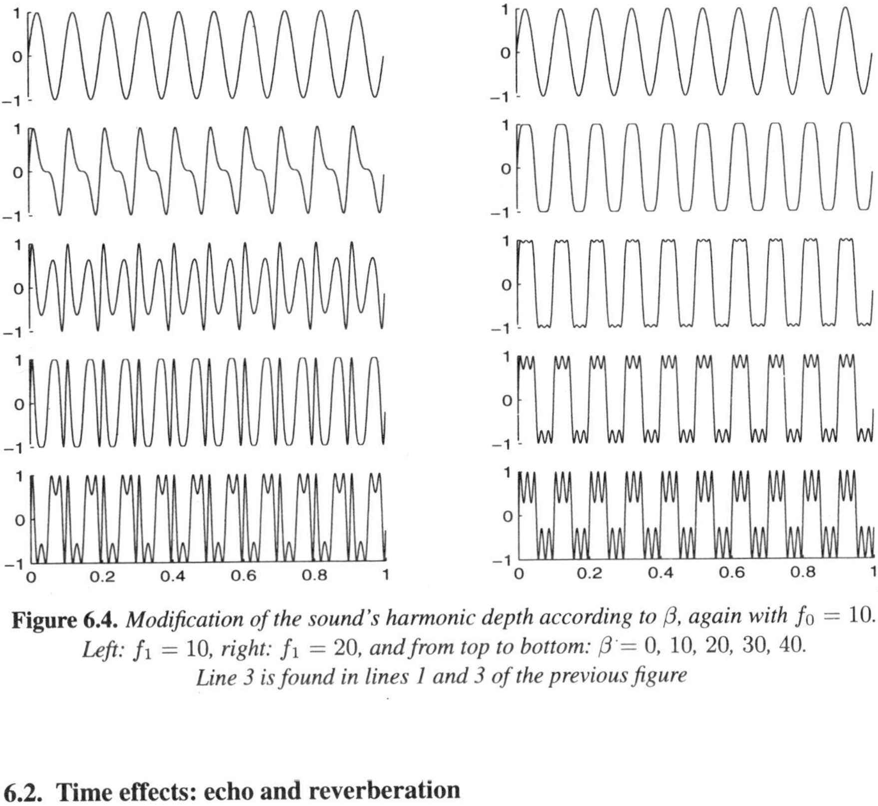

Figure 6.4. *Modification of the sound's harmonic depth according to β, again with $f_0 = 10$. Left: $f_1 = 10$, right: $f_1 = 20$, and from top to bottom: $\beta = 0$, 10, 20, 30, 40. Line 3 is found in lines 1 and 3 of the previous figure*

6.2. Time effects: echo and reverberation

These effects apply directly to the time signal. Here, we will be considering a discrete causal signal x, of finite *length* $N + 1$, originating from the sampling of a sound with frequency F_e. Therefore, we have $x_n = 0$ for $n < 0$ or $n > N$, and two consecutive samples are separated by the time interval $\tau = 1/F_e$.

6.2.1. *Simple echo*

Simple echo is the easiest effect to program. It consists of adding to the initial sound the same sound, with a *delay* r and attenuated by a factor $0 < g < 1$ called the *gain*. We will assume that r is an integer multiple of τ: $r = m\tau$. Hence the resulting signal y is such that

$$y_n = x_n + gx_{n-m}. \tag{6.2}$$

This is the echo that would be produced by the reflection of the sound on a perfectly reflecting wall located at a distance $d = cr/2$, with $g \simeq 1/(2d)$, if we assume that the source and the listener are placed at the same point.

COMMENT.– Using the formula (5.11) for the delay, relation (6.2) between x and y becomes, by Z-transform,

$$Y(z) = (1 + gz^{-m})X(z).$$

By considering the polynomial $b(z) = 1 + gz^m$, we thus have

$$Y(z) = b(z^{-1})X(z),$$

which means, because of (5.12), that y is the result of filtering x with the filter with the following Z-transform

$$H(z) = b(z^{-1}), \tag{6.3}$$

and, because of (5.10) with $v = b$, the samples of this filter are $h_0 = 1$, $h_m = g$, the other h_k being equal to zero. This is an FIR filter.

6.2.2. *Multiple echo*

Let us now examine the case of two walls facing each other, and generating a sequence of diminishing echoes: this is *multiple echo*. We will assume for example that the listener and the source are located close to one of the walls. After a first reflection on the opposite wall, the sound will return, then go back again to reflect a second time, then a third, and so on. Let y be the resulting sound. To account for the fact that the signal originating from the last reflection is no longer gx_{n-m} but instead gy_{n-m}, relation (6.2) becomes:

$$y_n = x_n + gy_{n-m}. \tag{6.4}$$

This is a *recurrence relation* that can be solved by hand, which leads to

$$y_n = x_n + gx_{n-m} + g^2x_{n-2m} + \ldots + g^kx_{n-km} + \ldots \tag{6.5}$$

This relation shows that the successive reflections are increasingly damped (since $0 < g < 1$), but it is useless computation-wise, because the number of necessary operations increases with n (roughly $3n/m$ operations to calculate y_n), whereas in (6.4), the computation of y_n requires at the most an addition and a multiplication.

COMMENT.– If we take the Z-transform of (6.4), which is also written $y_n - gy_{n-m} = x_n$, we now find

$$(1 - gz^{-m})Y(z) = X(z).$$

By considering the polynomial $a(z) = 1 - gz^m$, we thus have $a(z^{-1})Y(z) = X(z)$, which means, because of (5.12), that y is the result of filtering x with the filter with the following function as its Z-transform

$$H(z) = \frac{1}{a(z^{-1})}. \tag{6.6}$$

By applying (5.10) with

$$v(z) = \frac{1}{a(z)} = \frac{1}{1 - gz^m} = \sum_{n=0}^{+\infty} g^n z^{mn},$$

we infer that the samples of the filter h are $h_{mn} = g^n$, $n \geq 0$, the other h_k being equal to zero, where the k are non-multiples of m. This is an IIR filter. If we use (5.8) to compute $h * x$ with its coefficients, the result is the same as the one found by hand (6.5).

6.2.3. *Reverberation*

In a room with several walls, we will hear several reflections at different times, eventually all merging together to produce *reverberation*. In electronic music, artificial reverberation is obtained by passing the signal (converted temporarily into a mechanical signal) through several springs of different lengths: these are 'echo chambers' (first built by Hammond), found in most guitar amplificators. Designing a model for *digital* reverberation that does not require too much computation and, at the same time, is realistic, is something of an art, and the subject of many publications (see [DAF 02] and its bibliography). We will briefly describe two methods:

– the first is based on the room's impulse response, which must therefore be known, and requires a large number of operations;

– the second superposes simple and multiple echoes combined with all-pass filters. It requires less computation, but the results may not be as realistic.

6.2.3.1. *Using the impulse response*

We have already mentioned in section 1.5.2 that the sound y perceived by the listener is of the form

$$y = h * x,$$

where x was the sound produced by the orchestra and h is the room's impulse response. This was relevant to analog signals. It remains true for the associated sampled signals (see section 5.3.1), still denoted by x, h and y. Therefore, if the samples of the room's impulse response are available (they can be obtained by recording a brief sound close to the Dirac impulse), it is possible to obtain the discrete sound y by performing the convolution product of the sampled signals x and h.

If the room's response is not available, another option consists of imitating it using a random number generator [DAF 02]. The impulse response h below is obtained by filtering a noise w that decreases exponentially. In this case, b is a low-pass filter, the a_n are random numbers and d is a characteristic dimension of the room for which we wish to simulate reverberation:

$$h = b * w,$$

$$w_n = a_n \exp(-n\tau/d).$$

In both cases, computing the convolution $h*x$ requires a large number of additions and multiplications, making this method difficult to apply in real-time.

6.2.3.2. *Using echoes and all-pass filters*

This technique consists of combining the two kinds of echoes discussed previously. Figure 6.5 shows an example of a structure for simulating reverberation. In what follows, in order not to burden our notations, we will not write the indices of the parameters g, μ, ..., associated with the different elements of R and E, but understand that they *can vary* from one element to another.

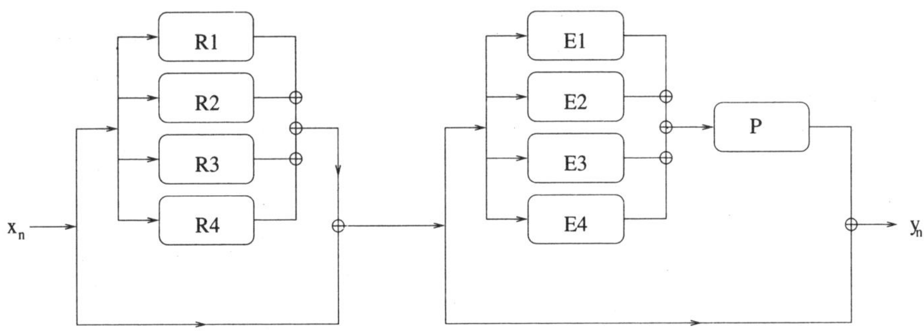

Figure 6.5. *Moorer's reverb, comprised of lines with a delay R, multiple echoes E and an all-pass filter P. The ⊕ represent additions*

The elements denoted by R are simply delays, their purpose being to simulate the first reflections. They each produce a signal of the form

$$u_n = gx_{n-m}$$

where $m = rF_e$, r being the delay that can vary for example from 20 to 100 ms for the set of elements R. The factor g is the gain of the simple echo, which can be set as $g = \mu/(cr)$, where $0 < \mu \leq 1$ represents the room's absorption, and cr is the distance covered by the sound to come back to its starting point after it has been reflected.

Elements denoted by E are multiple echoes combined with low-pass filters. When such an element E receives a signal u, it produces a signal

$$w = \mu h * v$$

where μ is still an absorption parameter, h is a low-pass filter (otherwise we would get a 'metallic' sound) and v verifies a version of (6.4):

$$v_n = u_{n-m} + g v_{n-m}.$$

Each gain g associated with an element E is chosen in the form $g = 10^{-3r/T_R}$, where $r = m\tau$ is the associated delay, chosen between 50 and 80 ms, and T_R is the *reverberation duration*, which by convention is the time it takes for a brief reverberated sound to decrease in intensity by 60 dB.

Finally, the last element, denoted by P, is an all-pass filter (see section 1.5.3). When it receives a signal w, it produces a signal y that obeys the recurrence equation

$$y_n = g y_{n-m} + w_{n-m} - g w_n. \tag{6.7}$$

The parameters of this filter suggested by Moorer [DAF 02] are $g = 0.7$ and $m = 6F_e 10^{-3}$ (still assumed to be an integer). This filter 'blurs everything together' by modifying the phases of each frequency differently, reproducing the effect a room has on sounds.

COMMENT.– Using the Z-transform of (6.7), we get, in this case

$$Y(z) = \frac{-g + z^{-m}}{1 - gz^{-m}} W(z),$$

and the Z-transform of the associated filter is

$$H(z) = \frac{-g + z^{-m}}{1 - gz^{-m}}.$$

Based on (5.13) this filter's transfer function is

$$H_t(f) = \frac{-g + \exp(-2mi\pi f/F_e)}{1 - g\exp(-2mi\pi f/F_e)}$$

which, because g is a real number, is such that $|H_t(f)| = 1$. Hence this filter is an all-pass filter: it does not modify the intensity of pure sounds, only their phases.

6.3. Effects based on spectrum modification

Other effects are based on a direct modification of the sound's spectrum, instead of time modifications, by modifying either the relative weights of the harmonics or partials, like the 'wah-wah' effect, or the frequencies themselves like the vibrato or the Leslie effect.

6.3.1. *The 'Wah-wah' effect*

The *wah-wah* effect brings to mind the formants of the vowels 'o' and 'ah', hence the name of course. It consists of adding to the initial sound the sound obtained by passing this initial sound through a band-pass filter with a band that *varies* with time: low for the 'o' sound and higher for the 'ah' sound (see Figure 1.29). Here is a digital description.

6.3.1.1. *An example of a band-pass filter*

A digital band-pass filter that requires few computations can be obtained using the recurrence formula

$$y_n = (1+c)(x_n - x_{n-2})/2 - d(1-c)y_{n-1} + cy_{n-2}, \tag{6.8}$$

the Z-transform of which is given by

$$H(z) = \frac{(1+c)(1-z^{-2})/2}{1+d(1-c)z^{-1} - cz^{-2}}. \tag{6.9}$$

Remember (see (5.13)) that the filter's transfer function is then

$$H_t(f) = H(\exp(2i\pi f/F_e)).$$

By choosing the parameters of this filter as follows:

$$c = \frac{\tan(\pi f_b/F_e) - 1}{\tan(\pi f_b/F_e) + 1},$$

$$d = -\cos(2\pi f_m/F_e),$$

we get a band-pass filter with a bandwidth f_b centered on f_m, hence with cut-off frequencies $[f_m - f_b/2,\ f_m + f_b/2]$. For a non-ideal filter, a cut-off frequency f_c is equivalent by convention to an *intensity divided by two* (hence -3 dB) compared to the maximum in the bandwidth (normalized to 1), and therefore equivalent to

$$|H_t(f_c)| = 1/\sqrt{2},$$

the intensity being proportional to the square of the amplitude, as we saw in Chapter 1. Inside the bandwidth, we have $|H_t(f)| \geq 1/\sqrt{2}$, whereas outside, $|H_t(f)| \leq 1/\sqrt{2}$. Figure 6.6 shows three examples of band-pass filters built based on relation (6.8).

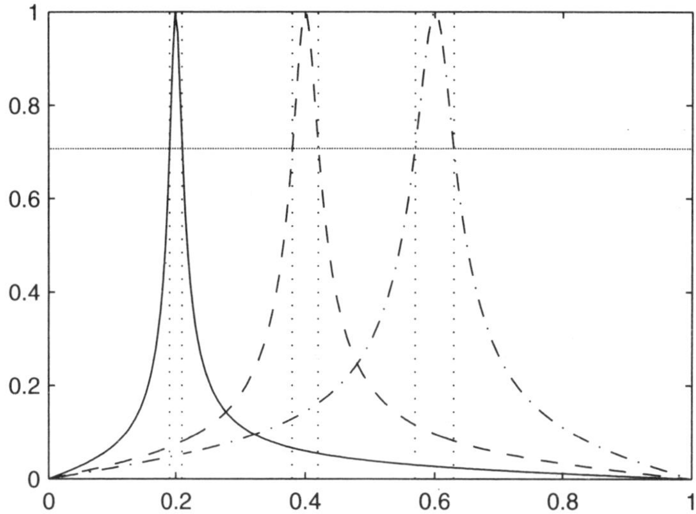

Figure 6.6. *Three band-pass filters corresponding to* $2f_m/F_e = 0.2, 0.4, 0.6$, *and* $f_b = f_m/10$. *The cut-off frequencies of the first filter are* $0.19F_e/2$ *and* $0.21F_e/2$

The *wah-wah* effect then consists of applying such a band-pass filter to the sound *by modifying with time the center frequency* f_m. The bandwidth can be maintained the same or not depending on your taste. Since f_m and f_b are constantly changing, the same goes for the coefficients c and d of recurrence relation (6.8) which becomes

$$y_n = (1 + c_n)(x_n - x_{n-2})/2 - d_n(1 - c_n)y_{n-1} + c_n y_{n-2},$$

with

$$c_n = \frac{\tan(\pi f_b(n)/F_e) - 1}{\tan(\pi f_b(n)/F_e) + 1}, \quad d_n = -\cos(2\pi f_m(n)/F_e).$$

In the examples of Figures 6.7 and 6.8, we chose $f_m(n)$ and $f_b(n)$ as follows:

$$f_m(n) = 2\,000 + 1\,000 \sin(2\pi n/F_e),$$

$$f_b(n) = f_m(n)/10.$$

Notice that the *wah-wah* effect does not modify the sound's pitch (this is clearly visible in Figure 6.8). It only changes the relative weights of the different frequency components, like the formants of the human voice. Another version of the *wah-wah* effect consists of setting up variable band-pass filters (or possibly band-stop filters) in parallel, each one affecting a different part of the spectrum.

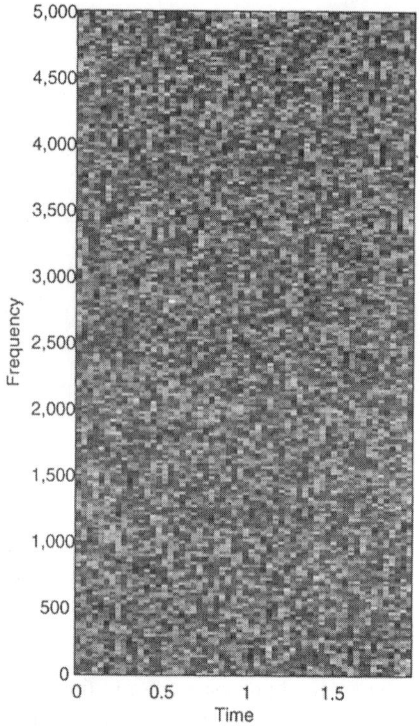

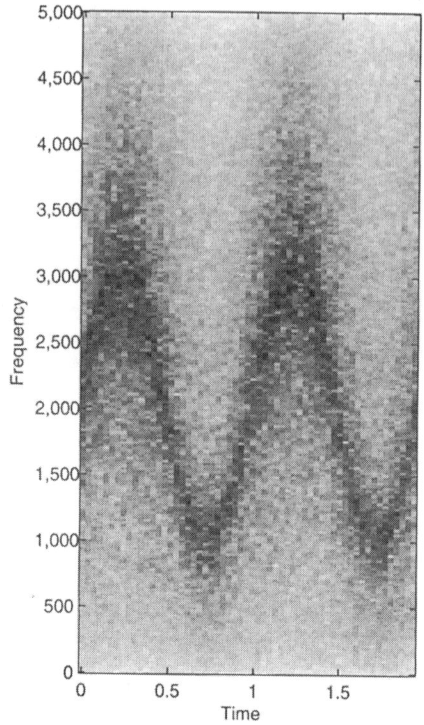

Figure 6.7. *Wah-wah effect (right) on white noise (left). The band-pass filter's center frequency f_m varies between 1,000 and 3,000 Hz, with a bandwidth $f_m/10$*

6.3.2. *AM or FM type sound effects*

These effects are based on a periodic variation of the amplitude:

$$s(t) = (1 + \eta \sin(2\pi f_1 t)) \sin(2\pi f t), \qquad (6.10)$$

or of the frequency:

$$s(t) = \sin(2\pi f t + \beta \sin(2\pi f_1 t)/f_1). \qquad (6.11)$$

The AM effect creates a beat with frequency f_1, something we have already encountered several times. The FM effect produces a vibrato with frequency f_1 and a frequency excursion β (see section 6.1.3). In the Leslie effect, the two are combined with a stereo effect.

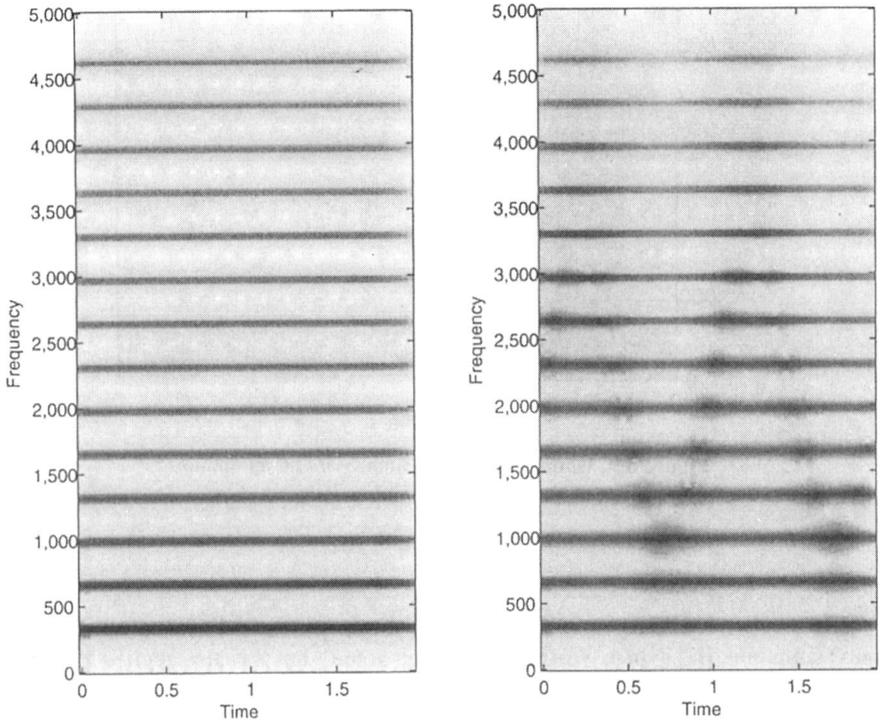

Figure 6.8. *Wah-wah effect (right) on a periodic sound (left). Same parameters as before*

6.3.2.1. *Vibrato*

The voice and many instruments produce a *vibrato*, the frequency of which varies between roughly 4 and 12 Hz. It is easy to produce such a vibrato with additive synthesis: we simply have to write each harmonic in the same way as in (6.11). Figure 6.9 represents the following sound, with fundamental 440 Hz, and a vibrato of 5 Hz:

$$s(t) = e(t) \sum_{n=1}^{9} \alpha_n \sin(880 n\pi t + 7\sqrt{n} \sin(10\pi t)).$$

The other parameters are $\alpha_n = 1\,000, 300, 0.01, 0.01, 0.01, 3, 0.5, 1, 4$ and $\beta = 7\sqrt{n}$. A single, trapezoid-shaped envelope $e(t)$ was used.

Aside from additive synthesis, such an effect, which modifies the sound's frequency, can be obtained by playing the data, or its digital equivalent, at a variable speed. But it cannot be achieved using filters, because the time invariance condition is not satisfied.

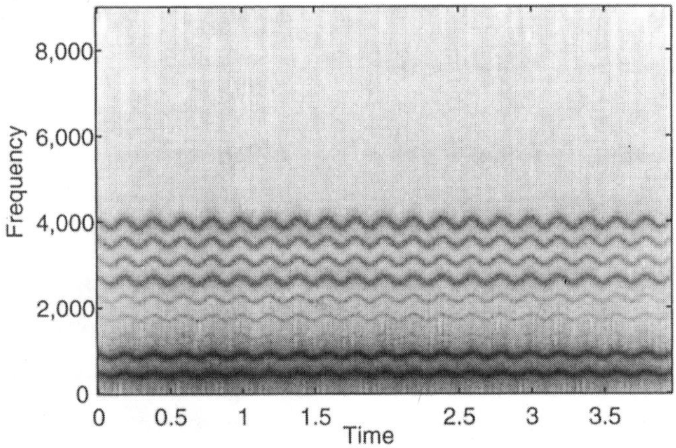

Figure 6.9. *Vibrato of a sound comprised of 9 harmonics. Does this remind you of anything?.*

6.3.2.2. *Leslie effect*

The *Leslie* effect was invented by Donald Leslie in the 1940's. It was integrated into Hammond, Baldwin or Wurlizer electronic organs, but was also applied to voices, for example in *Blue Jay Way* by the Beatles. It is achieved by two rotating loudspeakers facing opposite directions, producing a Doppler effect (see section 6.4.1) coupled with a variation of intensity.

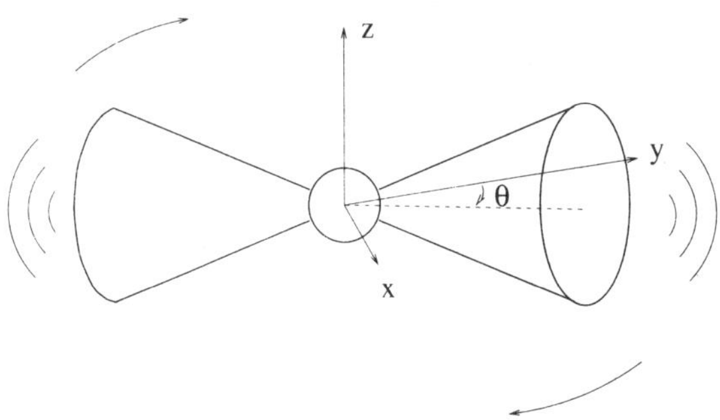

Figure 6.10. *Leslie effect produced by two rotating speakers. It became popular with the use of electric organs*

This effect, which is inevitably stereo, consists of sinusoidal variations of the amplitude and the frequency, with the left and right channels in opposite phases, the variations of the amplitude and the frequency being in quadrature, as indicated in Figure 6.11 for an entire rotation of the device. The rotation speed is roughly 3 to 6 rounds/s.

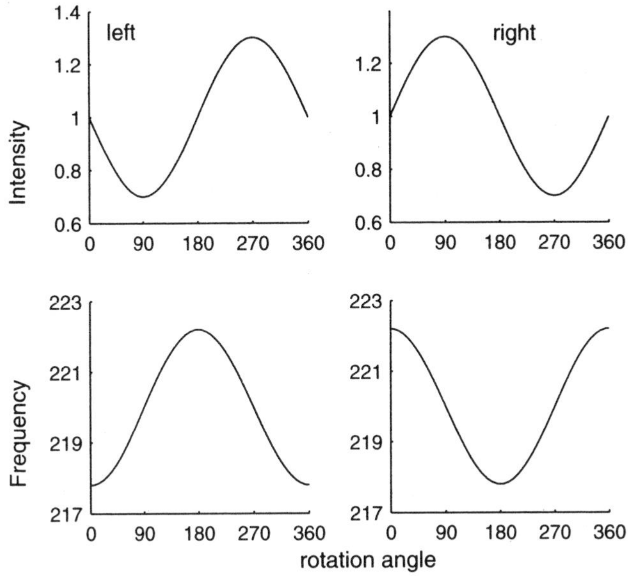

Figure 6.11. *Leslie effect: amplitude (top) and frequency (bottom) variations of each channel, over a complete rotation of the loudspeakers*

Such an effect can be implemented by additive synthesis of the left and right channels $s_g(t)$ and $s_d(t)$. This is done by writing each harmonic in amplitude or frequency modulated form:

$$s_g(t) = \sum_{n \geq 1} e_n(t)(1 - \eta \sin(2\pi f_L t)) \sin(nf(2\pi t - \mu \sin(2\pi f_L t)/f_L)),$$

$$s_d(t) = \sum_{n \geq 1} e_n(t)(1 + \eta \sin(2\pi f_L t)) \sin(nf(2\pi t + \mu \sin(2\pi f_L t)/f_L)),$$

where f is the frequency of the fundamental, $e_n(t)$ is the envelope of each harmonic, and for example $\eta = 0.3$, $\mu = 0.01$ and $f_L = 4$. Note that the instantaneous frequencies are (by overlooking the effect produced by the amplitude variation)

$$f_{\text{inst}(g,n)} = nf(1 - \mu \cos(2\pi f_L t)),$$

$$f_{\text{inst}(d,n)} = nf(1 + \mu \cos(2\pi f_L t)).$$

The spectrograms of each channel are shown in Figure 6.12. For clarity, the values of μ and η have been exaggerated ($\eta = 1$ and $\mu = 0.1$).

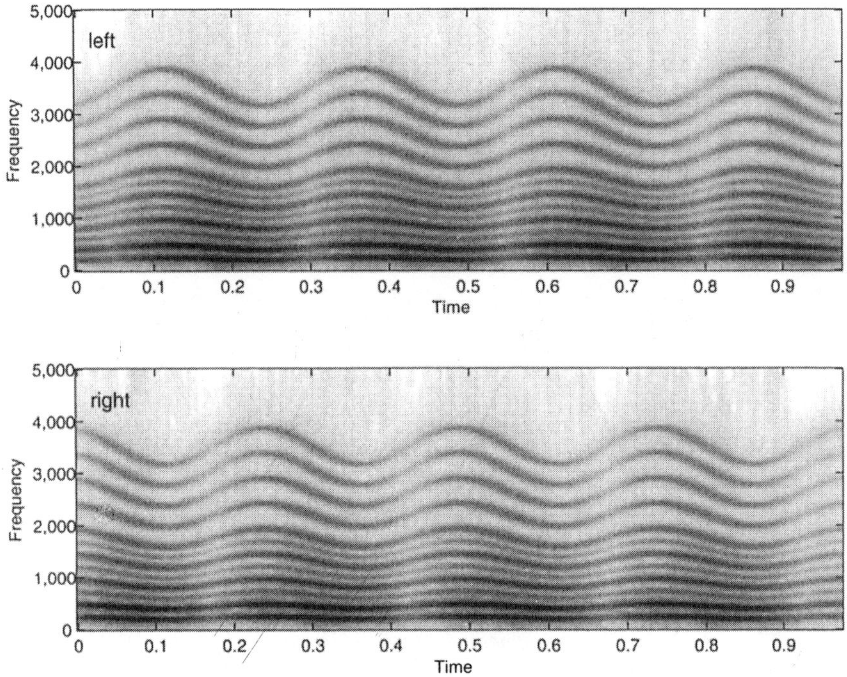

Figure 6.12. *Spectrograms of the Leslie effect (we overdid it a little...)*

6.4. Study problems

6.4.1. *The Doppler effect (**)*

A harmonic punctual sound source with frequency f, located at the origin, generates a pressure wave of the form

$$p(x, y, z, t) = \frac{\exp(ikr - 2i\pi ft)}{r}$$

with $r = \sqrt{x^2 + y^2 + z^2}$. In both of the following cases, calculate and plot the instantaneous frequency perceived by a listener:

– in uniform rectilinear motion (therefore its position is of the form $M(t) = M_0 + tV$, $V = (V_1, V_2, V_3)$ being the speed vector and M_0 its position at instant $t = 0$);

– moving along a circle at a constant angular speed.

In the first case, proceed by establishing the limits of the instantaneous frequencies for $t \to \pm\infty$, and study also the particular case where the vectors M_0 and V are parallel.

6.4.2. *FM and Chowning (***)*

Let s be the following frequency-modulated sound

$$s(t) = \sin(2\pi ft + \beta \sin(2\pi gt)/g).$$

We are going to conduct the frequency analysis of this sound, and determine the interesting cases where it is periodic. We will be using the two following trigonometry formulas:

$$\sin(a+b) = \sin a \cos b + \sin b \cos a,$$
$$2 \sin a \cos b = \sin(a+b) + \sin(a-b).$$

1) Let $\mu = \beta/g$. Check that

$$s(t) = \sin(2\pi ft)\cos(\mu \sin(2\pi gt)) + \cos(2\pi ft)\sin(\mu \sin(2\pi gt)).$$

Notice that

$$\exp(i\mu \sin(2\pi gt)) = \cos(\mu \sin(2\pi gt)) + i \sin(\mu \sin(2\pi gt)).$$

The Bessel function of the first kind (seen before when we discussed membranes) can be expressed as

$$J_n(\mu) = \frac{1}{2\pi}\int_0^{2\pi} \exp(i(\mu \sin x - nx))\, dx. \tag{6.12}$$

Show that the Fourier series expansion of $\exp(i\mu \sin(2\pi gt))$ is as follows:

$$\exp(i\mu \sin(2\pi gt)) = \sum_{n=-\infty}^{+\infty} c_n \exp(2i\pi ngt)$$

with

$$c_n = J_n(\mu).$$

2) Check that the values $J_n(\mu)$ are real numbers, then, by making the variable change $y = \pi - x$ in (6.12), check that $J_n(\mu) = (-1)^n J_{-n}(\mu)$. Use this result to show that

$$\cos(\mu \sin(2\pi gt)) = c_0 + 2\sum_{n\geq 1} c_{2n} \cos(4\pi ngt),$$

$$\sin(\mu \sin(2\pi gt)) = 2\sum_{n\geq 0} c_{2n+1} \sin(2\pi(2n+1)gt),$$

and, finally, show that

$$s(t) = J_0(\mu)\sin(2\pi ft) + \sum_{n\geq 1} J_n(\mu)\left[\sin(2\pi(f+ng)t) + (-1)^n \sin(2\pi(f-ng)t)\right].$$

3) We will ignore the fact that some of the values $J_n(\mu)$ may be equal to zero. What can you say about $s(t)$ – is this sound periodic? If it is, what is its fundamental? What are the harmonics present? – in the following cases:

– $g = f$;

– $g = f/q$ with $q > 0$ an integer;

– $g = pf/q$ with $p > 0$, $q > 0$ two integers with no common divisors (answer: the fundamental is f/q).

6.5. Practical computer applications

6.5.1. *Sound synthesis*

Record and analyze (time, frequency and time-frequency) a note from an instrument of your choice. Using this analysis, create a function based on the same idea as the `instrument` function written for the applications in Chapter 2 and that imitates as best as possible the sound that you recorded.

6.5.2. *Chowning synthesis*

Still based on the same idea as the `instrument` function, create a function called `chowning` with the header

$$\texttt{function s} = \texttt{chowning(f0, T, Fe)}$$

that delivers a sound of the form

$$s(t) = \sin(2\pi f_0 t + \beta \sin(2\pi f_1 t)/f_1).$$

The values of f_1 and β will be specified inside the function itself. This way, we keep the general format of the instrument functions, which allows us to play `chowning` using the `play` function without having to modify it. Try different values.

To create a stereo sound, you can create a left channel `sl` and a right channel `sr` based on the previous model, then combine them in a single matrix that can be read by the `sound` function. The commands are as follows:

computation of `sl`
```
f0 = f0+0.8;
```
computation of `sr`
```
s = [sl;sr]';
```
Note the slight change in frequency before going on to computing `sr`, so as to produce a phase effect. Explanation of the last line: `sl` and `sr` are two line vectors (that is how they should have been programmed), and the `;` that separates them indicates that they

are arranged one below the other in a matrix with two lines and as many columns as the number of elements of `sl`. This implies that `sl` and `sr` must have the same number of elements. Finally, the 'prime' transposes this matrix so that it may be recognized by the `sound` function in the case of stereo sounds.

6.5.3. *Reverberation*

Moorer's reverb is programmed in the `reverb.m` file found on the `AM` website. Copy it to your working directory, read it and make the connection with the description made in this chapter. This will allow you to make modifications to the different parameters (delay, reverberation time, etc.). Use this function inside one of your `instrument` functions to modify its sonority.

6.5.4. *Vibrato*

Use the formula written in this chapter to add vibrato to your instrument. The most convenient way to do this is to modify the `synthad.m` function: save it under a different name, `synthadv.m` for example, and make the modifications inside this new file. To add realism, you can start the vibrato half a second after the beginning of the sound.

6.5.5. *The Leslie effect*

Open a new file based on the same idea as `instrument.m`, called `leslie.m`, and inside which you will program the sound of an organ (remember that the typical organ note is composed of several pipes, the fundamentals of which follow a harmonic progression, typically 1, 2, 3, 4, 5). Inside this function, invoke a new function `synthadl.m` (based on the `synthad.m` model) that needs to be created and will provide as its output a stereo sound (see the section on the Chowning synthesis for the format of such a sound), and inside which each 'harmonic' will be programmed based on the model described in this chapter. The envelope will be programmed inside the `synthadl.m` function itself. Then run the `play` program to play your new instrument.

Bibliography

[BAI 01] BAILHACHE P., *Une histoire de l'acoustique musicale*, CNRS Editions, Paris, 2001.

[COL 98] *Colloque acoustique et instruments anciens, factures, musiques et science*, Musée de la musique, 1998.

[COR 82] CORDIER S., *Piano bien tempéré et justesse orchestrale*, Buchet/Chastel, Paris, 1982.

[DAF 02] *DAFX – Digital Audio Effects*, John Wiley and Sons, New York, 2002.

[ESC 01] ESCUDIÉ B., GAZANHES C., TACHOIRE H., TORRA V., *Des cordes aux ondelettes*, Publications de l'Université de Provence, Aix-en-Provence, 2001.

[FIC 96] FICHET L., *Les théories scientifiques de la musique*, Librairie J. Vrin, Paris, 1996.

[FLE 29] FLETCHER H., *Speech and Hearing in Communication*, Editions Van Nostrand, New York, 1929.

[FLE 98] FLETCHER N., ROSSING T., *The Physics of Musical Instruments*, Springer-Verlag, New York, 1998.

[GAS 90] GASQUET C., WITOMSKI P., *Analyse de Fourier et applications*, Masson, Paris, 1990.

[HEL 68] HELMHOLTZ H., *Théorie physiologique de la musique*, Masson et Fils, Paris, 1868.

[HON 76] HONEGGER M., *Science de la musique, dictionnaire de la musique*, Bordas, Paris, 1976.

[INS 95] *Les instruments de l'orchestre*, Bibliothèque Pour la science, Belin, Paris, 1995.

[JAR 71] JARGY S., *La musique arabe*, Collection Que sais-je, PUF, Paris, 1971.

[JOU 00] JOUHANEAU J., *Notions élémentaires d'acoustique, électroacoustique*, Editions TEC & DOC, Paris, 2000.

[JUN 79] JUNGHAUS H., *Der Piano und Flugelbau*, Verlag Das Musikinstrument, Frankfurt, 1979.

[LAS 94] LASCAUX P., THÉODOR R., *Analyse numérique matricielle appliquée à l'art de l'ingénieur*, Masson, Paris, 1994.

[LEI 80] LEIPP E., *Acoustique et musique*, Masson, Paris, 1980.

[MAT 69] MATHEWS M., *The Technology of Computer Music*, MIT Press, 1969.

[MOR 95] MOREAU N., *Techniques de compression des signaux*, Masson, Paris, 1995.

[PIE 99] PIERCE J., *Le son musical*, Bibliothèque Pour la science, Belin, Paris, 1999.

[PRO 95] *Proceedings of the International Symposium on Musical Acoustics*, Le Normont, Dourdan, France, Editions SFA, July 1995.

[RAM 37] RAMEAU J.-P., *Génération harmonique*, Prault fils, Paris, 1737.

[RIS 99] RISSET J., 'Hauteur, harmonie, timbre, synthèse', *Musique, rationalité, langage – l'harmonie : du monde au matériau*, L'Harmattan, Paris-Montréal, p. 153-167, 1999.

[SCH 66] SCHAEFFER P., *Traité des objets musicaux*, Le Seuil, Paris, 1966.

[SHA 49] SHANNON C., *The Mathematical Theory of Communication*, University of Illinois Press, 1949.

[SON 79] *Sons et musique*, Bibliothèque Pour la science, Belin, Paris, 1979.

[UNI 80] *Encyclopédie Universalis*, Encyclopédie Universalis France, Paris, 1980.

[VAL 93] VALETTE C., CUESTA C., *Mécanique de la corde vibrante*, Hermès, Paris, 1993.

[ZWI 81] ZWICKER E., FELDTKELLER R., *Psychoacoustique, l'oreille récepteur d'information*, Masson, Paris, 1981.

Index

V

W

X

Y

Z